**MEDICAL
INTELLIGENCE
UNIT 18**

Necrotizing Enterocolitis

Brian F. Gilchrist, M.D., F.A.C.S.
The State University of New York (SUNY)
Health Science Center
Assistant Professor of Pediatric
Surgery and Transplantation
Brooklyn, New York, U.S.A.
and
Director of Pediatric Surgery
Staten Island University Hospital
Staten Island, New York, U.S.A.

CRC Press
Taylor & Francis Group
Boca Raton London New York

CRC Press is an imprint of the
Taylor & Francis Group, an **informa** business

NECROTIZING ENTEROCOLITIS

Medical Intelligence Unit
Designed by Judith Kemper

First published 2000 by Landes Bioscience

Published 2018 by CRC Press
Taylor & Francis Group
6000 Broken Sound Parkway NW, Suite 300
Boca Raton, FL 33487-2742

© 2000 by Taylor & Francis Group, LLC
CRC Press is an imprint of Taylor & Francis Group, an Informa business

First issued in paperback 2019

No claim to original U.S. Government works

ISBN 13: 978-0-367-44733-5 (pbk)
ISBN 13: 978-1-58706-008-3 (hbk)

Visit the Taylor & Francis Web site at
http://www.taylorandfrancis.com

and the CRC Press Web site at
http://www.crcpress.com

Library of Congress Cataloging-in-Publication Data

CIP applied for but not received at time of publication.

DEDICATION

This work is dedicated to Max L. Ramenofsky, M.D., who has taught us by example the manly dictum of understanding. "Never judge another man unless you have walked a hundred miles in his moccasins." Dr. Ramenofsky possesses a spirit of independence that is both brave and pure, he accepts man as man. He allows a man to be a man and in so doing taps a man's greatest potential.

Brian F. Gilchrist, M.D., F.A.C.S.
Brooklyn, New York, U.S.A.
1999

CONTENTS

EDITOR

Brian F. Gilchrist, M.D., F.A.C.S.
The State University of New York (SUNY)
Health Science Center
Assistant Professor of Pediatric
Surgery and Transplantation
Brooklyn, New York, U.S.A.
and
Director of Pediatric Surgery
Staten Island University Hospital
Staten Island, New York, U.S.A.
Chapter 10

CONTRIBUTORS

Colin A.I. Bethel, M.D.
Assistant Professor of Surgery
UMDNJ
Newark, New Jersey, U.S.A.
Chapter 6

Christopher Breuer, M.D.
Instructor in Pediatric Surgery
Brown University
Hasbro Children's Hospital
Providence, Rhode Island, U.S.A.
Chapter 9

Benjamin Z. Cooper, M.D.
Instructor in Surgery
SUNY-HSCB
Brooklyn, New York, U.S.A.
Chapter 2

Carlotta Hample, M.D.
Assistant Professor of Pediatrics
SUNY-HSCB
Brooklyn, New York, U.S.A.
Chapter 8

Katrine Hasen, M.D.
Assistant Professor of Pathology
Brown University
Hasbro Children's Hospital
Providence, Rhode Island, U.S.A.
Chapter 7

Marc Lessin, M.D.
Assistant Professor of Surgery
Tufts University
The Floating Hospital
Boston, Massachusetts, U.S.A.
Chapter 5

Steven Piecuch, M.D.
Clinical Assistant Professor
Attending Neonatologist
SUNY-HSCB, Department of Pediatrics
Brooklyn, New York, U.S.A.
Chapter 3

John N. Schullinger, M.D., F.A.C.S.
Emeritus Professor of Surgery
Babies Hospital
  of New York
Columbia-Presbyterian Medical Center
New York, New York, U.S.A.
Chapter 1

Richard J. Scriven, M.D.
Instructor in Surgery
SUNY-HSCB
Brooklyn, New York, U.S.A.
Chapter 2

E. Christine Wallace, M.B.
Assistant Professor of Radiology
Tufts University
The Floating Hospital
Boston, Massachusetts, U.S.A.
Chapter 4

PREFACE

I first met Necrotizing Enterocolitis (NEC) in Memphis among the 125 bassinets at the University of Tennessee's Newborn Center. Thom Lobe made a most insightful comment about the disease when he said that there must be a common denominator that caused it. Thom found the disease fascinating, yet frustrating. He and others with truly fine and visionary minds, however, have pursued less pedestrian horizons. While Adzick and Crombieholme and Luks studied fetal surgery, and Lobe and Georgeson and Holcombe laparoscopically revolutionized pediatric surgery, NEC lay there as an orphan.

I was often struck by the disparate outcomes of the infants insulted by NEC. Why, I queried, did one baby, with all other factors being equal, die within 24 hours, while another survived with a 10-day course of medical care, while still another baby fought, was resected, medically treated and survived. The literature on NEC, especially espoused in the major textbooks, merely recapitulated what had been written in the past; there has been no new guidance. NEC was viewed as a *fait accompli*; a sometimes-dire disease that was to be confronted only after it reared its perfidious head. All one has to do is read Amoury's 1980 chapter in Ashcraft and compare it to Kosloske's 1996 rendition in Columbani to see that philosophically our approach had not changed in a generation.

It was during my time with Dr. Monaco at Harvard that I realized that this disease needed to be viewed from a radically different perspective. He emphasized that the disease had to be addressed before it became apparent, a concept he had gleaned fighting organ rejection for 30 years. I took away from Boston a radical exhortation, as did our forebears at Faniuel Hall; a paradigm shift was needed to address NEC. This monograph asks the readers to rethink their approach to NEC. Although, there are no ultimate answers yet, no Jesuitical certainty, we have put together a work that I hope will compel others to pursue a way to intervene before the die has been cast. How sad for all of us, if 30 years from now, a student exclaims with incredulity: "Can you believe that they didn't know how to block and treat NEC back then?" I remember my grandfather and father lamenting cases they lost for want of antibiotics in Springfield, Massachusetts in the 1930s and 1940s. Their sorrow was still palpable 20 years later, a sorrow born of the frustration of not having the weapons needed for the fight. We must look for a unifying concept, most likely an immunologic one, which will allow us to intervene in this disease in a prophylactic manner. We must be forewarned and armed.

Thus, this book is a promissory note and a challenge. The challenge is to resist being comfortable. The authors whom you will meet on these pages make up an eclectic team that I believe will bring you to a new way of thinking about NEC. The group consists of veterans and rookies, men and women, surgeons and nonsurgeons. Each of the writers has spent wearying hours attempting to diagnose and treat NEC. The authors are investigators and clinical doctors with the nobility attendant with being such. Each believes that the paradigm for the treatment of NEC must be changed and each is infused with John Kennedy's plea: "Although some children may be the victims of fate, none will be the victims of our neglect." We humbly offer you our thoughts.

Brian F. Gilchrist, M.D., F.A.C.S.
Brooklyn, New York, U.S.A.
1999

"The heights by men reached and kept
were not attained by sudden flight
But they, while their companions slept,
Were toiling upward in the night."

— *Henry Wadsworth Longfellow*
The Ladder of St. Augustine,
Stanza 10

CHAPTER 1

The History of Necrotizing Enterocolitis Investigation

John N. Schullinger

Neonatal necrotizing enterocolitis (NEC) has been recognized in the United States for over 30 years. Prior to this time, infants with NEC were most likely included among sporadic cases reported as neonatal perforations in which various etiologies were proposed.

Credit for first reporting NEC has been given to Generisch[1] (1891), who described a 45-hour-old premature infant with vomiting and abdominal distention. The baby died within 24 hours from a perforation in the ileum without mechanical obstruction. However, a detailed description of the postmortem findings, as pointed out by Rowe,[2] was consistent with meconium peritonitis. Rowe suggested, instead, that the five newborns described three years previously (1888) by Paltauf[3] had clinical courses and postmortem findings more consistent with NEC and should be regarded as the first recorded cases.

Following Thelander's 1939 report[4] on a series of 85 newborns with nonobstructive gastrointestinal perforations dating back to 1825, little attention was focussed on this topic for 25 years. Intermittent reports continued to appear in the European and American literature, but it was not until the early 1960s that the disease began to be regarded in this country as a distinct clinicopathologic entity. How this came into being is an interesting story and worth the telling.[5]

It was common practice at the Babies Hospital in New York City to send the surgical specimens of the day to the pathology laboratory late in the afternoon or evening. Since they were not processed until the following day, a certain amount of patchy autolysis was an expected finding, especially in patients with severe enteritis and peritonitis. One day in the early 1950s, Dr. William A. Blanc, the Chief Pathologist at Babies Hospital, received from Dr. Thomas V. Santulli, Chief of the Pediatric Surgical Service, a fresh specimen of intestine from an infant operated upon for a fulminating enterocolitis. Upon examination of the specimen Dr. Blanc was surprised to find that its necrotic appearance, gross and microscopic, was exactly what he would have expected to find in the usual 12 to 24 hour old specimen; yet, here it was in one that had been freshly excised. He realized that what he was observing was not the customary postmortem autolysis, but the disease itself. From this observation the clinical entity of acute necrotizing enterocolitis* came into being.

---

* The use of this term had first appeared in the European literature in 1953 in reports by Schmid[6] and Quiser[7] on pathologic and clinical studies of a series of infants dying from necrotic lesions of the gastrointestinal tract.

Necrotizing Enterocolitis, edited by Brian F. Gilchrist. ©2000 Eurekah.com.

Following a report on six infants with colonic perforation due to necrotizing colitis,[8] two comprehensive reports on NEC appeared from the Babies Hospital in the mid 1960s. The first significant surgical experience from the same institution was published in 1967.

In 1964, Berdon[9] described 21 cases of NEC from the premature nursery of the Babies Hospital seen between 1954 and 1964. There were four survivors, three of whom underwent operations. The clinical features, including the radiographic signs of pneumatosis and portal vein gas, are described along with the surgical and postmortem findings. The authors speculated on the role of bacteria and a possible localized Schwartzman-type phenomenon in the etiology of the disease. A year later Mizrahi[10] made a detailed study of this same group of patients and noted the frequent isolation of gram negative organisms in both ante- and postmortem specimens. He suggested that the disorder might be due to a deficiency of lysozyme in cow's milk formulae or to the direct action of gram negative bacterial endotoxins in the presence of increasing amounts of stress released catecholamines.

Following these reports, interest in NEC spread rapidly throughout the country. More authors[11,12] began to publish on their experiences with the disease. Early investigations developed along two main lines:

1. laboratory and clinical studies of pathogenesis and
2. clinical studies directed towards management. Issues of prevention were also addressed in the course of these studies.

Pathogenesis

As a result of these early clinical and experimental investigations, it became apparent that three essential factors were important in the development of NEC:

1. mucosal injury,
2. intraluminal metabolic substrates, and
3. the presence of bacteria.[13] Added to these was the recognition of certain neonatal physiologic and immune characteristics which were thought also to contribute to the development or prevention of NEC, especially in the premature infant.

Indirect Mucosal Injury

Touloukian[14] reported on 25 cases of NEC in 1967. He called attention to the work of Scholander,[15] Johansen,[16] and others[17,18] who showed that in diving mammals and birds a physiologic and protective mechanism exists whereby in the presence of hypoxia there is a reflex bradycardia and redistribution in regional blood flow away from the splanchnic circulation to the heart and head. Based on clinical and pathologic observations in their series, Touloukian[14] postulated that activation of this protective reflex, brought about by asphyxia or shock in the perinatal period, could lead to shunting of blood away from the mesenteric, renal, and peripheral vascular beds. This course of events, he reasoned, results in areas of intestinal mucosal ischemia with breakdown in mucosal integrity, bacterial invasion, and death from either septicemia or perforation. Lloyd[19] also invoked this reflex in his report in 1969 on 87 newborns with gastrointestinal perforations in which 70 of his infants had experienced a significant hypotensive or anoxic insult.

Touloukian pursued this theory of pathogenesis in the laboratory, and in 1972 he reported on selective gut mucosal ischemia in asphyxiated neonatal piglets.[20] It was shown in this study that the total and isolated mucosal perfusion of the duodenal, proximal, and distal jejunal segments were significantly greater than that to the stomach, distal ileum and colon. Pathologically, the mucosa and submucosa of the asphyxiated and resuscitated piglets showed vascular congestion and hemorrhage consistent with the early lesion of NEC.

Barlow[21] carried Touloukian's work one step further. She demonstrated that formula-fed newborn rats subjected to hypoxic stress with or without gram negative bacterial contamination all developed the clinical and pathologic characteristics of NEC. The animals which had suckled on their mothers did not acquire the disease. Gamma globulin given either subcutaneously at birth or orally daily had a 60% protective effect in the formula fed rats as did Lactinex given daily. Oral neomycin had no protective effect. Further studies by Pitt[22] demonstrated the significance of breast milk leukocytes in the protective effect of breast milk in rats. Pitt found that frozen and thawed rat milk was protective only if leukocytes from fresh rat milk were added.

Necrotizing enterocolitis following umbilical vein exchange transfusion was documented in early reports by Hermann[23] (1965), Corkey[24] (1968), and Castor[25] (1968). Touloukian[26] in 1974 demonstrated a significant rise in portal venous pressure during the injection phase of exchange transfusion via the umbilical vein in newborn piglets. The possibility of mucosal ischemia secondary to venous congestion was raised.

Umbilical artery catheterization has also been implicated as a causative factor in NEC, especially where use has been prolonged or as a portal for infusion of medications or fluids in hemodynamically unstable neonates. Emboli to mesenteric vasculature have been reported, and recent studies using duplex Doppler sonography strongly suggest that these catheters can cause a decrease in mesenteric blood flow.[27]

Leake[28] was the first to recognize the association of hyperviscosity syndrome associated with NEC. He reported a full term infant with a peripheral hematocrit of 83% at 11 hours of life who at 19 hours shows widespread pneumatosis and died at 52 hours of NEC. Leake[28] postulated that hyperviscosity with poor microcirculation in the mesenteric vascular bed might have led to hypoxic tissue changes in the bowel wall. Subsequently Hankanson and Oh[29] reported 5 of 14 SGA infants with hyperviscosity who developed NEC, and then Le Blanc[30] demonstrated that NEC can be caused by producing polycythemic hyperviscosity in the newborn dog.

Other causes of circulatory mucosal ischemia have been well documented and include shock, hypothermia, congenital heart disease (i.e., patent ductus arteriosus and hypoplastic left heart syndrome), congestive failure, and more recently, aminophylline, indomethacin, and maternal use of cocaine.

Direct Mucosal Injury

Numerous clinical studies have identified a host of factors causing direct mucosal injury, including hyperosmolar formulas and medications, prolonged, intractable diarrhea, and various microbial organisms.

An increased incidence of NEC has been observed in low-birth-weight infants fed hyperosmolar elemental formulas.[31] In 1974 the mucosal lesions of NEC were produced experimentally in newborn goats fed hyperosmolar mother's milk, but not in those receiving unmodified mother's milk.[32] Hyperosmolar oral medications and solutions, such as undiluted calcium lactate and vitamin E, have also been associated with NEC. The mechanisms by which these substances injure the gut mucosa are unknown, but redistribution of blood supply and/or alteration of bowel flora resulting from a sucrose bolus are possibilities.[33]

In Polin's report[34] of 13 term infants with NEC, eight had diarrhea for at least 11 days prior to development of NEC, the diagnosis being made between the 26th and 67th day of life. He postulated that the mucosal damage could be either direct or the result of malnutrition associated with protracted diarrhea.

The possibility of bacteria causing direct mucosal injury was proposed by Waldhausen[8] in 1963, Kliegman[35] in 1979 and Lawrence[36] in 1982. This mechanism may explain the delayed onset of NEC in infants lacking other identifiable risk factors. *Clostridium perfringens, difficile,* and *butyricum* are particularly suspect in this regard.[35]

Intraluminal Substrates

The importance of feeding as a causative factor in NEC was recognized by Santulli[13] in our report on 64 cases seen over a 20 year period. All but one of these infants had been fed either dextrose solution or formula prior to the onset of symptoms. It was thought that these substrates provided the medium for the colonization of bacteria which in the presence of damaged mucosa could invade the bowel wall, proliferate, and initiate the series of pathologic events characteristic of NEC.

Support for this theory was found in the experiments in newborn rats by Barlow et al[21] described earlier in this Chapter. Untreated rat breast milk was found to be protective against NEC following ischemic mucosal injury, whereas simulated rat milk was not. Knowing that premature infants, like newborn rats, are relatively immune deficient with the degree of deficiency being dependent upon gestational age, early speculation on the prevention of NEC by breast milk feedings was expectedly enthusiastic. Aside from its high lactose, low protein composition, breast milk was known to contain many protective factors including secretory IgA, IgG and IgM, bifidus enhancing growth factor, antistaphylococcal factor, lactoperoxidase, interferon, anti-trypsin, T and B lymphocytes, neutrophils, and macrophages which are capable of producing complement, lysozyme, and lactoferrin. Breast milk appeared to be ideally constituted to provide protection against NEC.

Unfortunately, the expectations from breast milk feedings were not fully realized. The process of refrigerating, freezing, or pasteurizing interferes with the immunologic contribution of breast milk, including a diminished number of viable cells and levels of IgA. Many reports have documented the occurrence of NEC in infants fed exclusively on stored breast milk. However, it has been noted that when NEC does occur in these infants, the clinical course may be less fulminating.[35,37] Also, in a recent prospective study from England[38] the incidence of NEC was six times higher in formula fed infants than in those fed breast milk only.

Other investigations into the relationships of feeding and NEC centered around clinical studies involving volumes and rates of feeding. In 1978 Brown and Sweet[39] reported the virtual disappearance of NEC from their nursery during four years of a slowly progressive feeding regimen, and Goldman[40] in 1980 observed a decrease in the high incidence of NEC when feeding volumes were reduced and large volume increases discontinued. However, Book[41] stressed the importance of supplying adequate calories and protein to the premature infant and suggested on the basis of a prospective study that a rapid rate feeding schedule of 20 ml/kg daily increases in formula is not of major importance in the development of NEC.

Bacteria

The importance of bacteria in neonatal enteritis was first suggested by Blanc[42] who in 1959 reported the finding of purulent amniotic exudate in the lumen of the GI tract in infants born to mothers with amnionitis. He demonstrated that the swallowed amniotic fluid could produce enteritis in the newborn.

Early clinical studies recognized the contribution of bacteria in the pathogenesis of NEC and the wide range of organisms involved.[13] Prominent among those associated with NEC have been the hydrogen producing *Klebsiella* and *E. coli, Clostridium, Enterobacter, Pseudomonas, Staphylococcus, Candida,* and several viruses. Although, as previously stated, it had been proposed that NEC could be caused by abnormal colonization and direct injury to the mucosa by bacterial toxins, most students of the disease believe that bacteria are part of a series of events involving both ischemic mucosal injury and a suitable substrate.

Although most cases of NEC are sporadic, epidemics caused by single identified organisms have been reported. Involved organisms have included *Klebsiella, Salmonella, Pseudomonas,* and especially *Clostridium perfringens* and *butyricum.*[35] Small clusters have generally been associated with no single etiologic organism. The bacteriology of NEC has been the focus of much study and is presented in detail in Chapter two.

Conclusion

Why the disease begins and progresses in a single individual has yet to be fully elucidated. Recent studies of the effects of reperfusion injury, cytotoxic oxygen-derived free radicals, platelet activating factor, tumor necrosis factor, interleukins, and bacterial endotoxins have contributed greatly to an understanding of the mechanisms involved.[43]

However, the fact that the incidence of NEC in the United States ranges from only 1% to 7% is probably best explained by what Kosloske[44] has called the quantitative extremes of mucosal injury, bacterial pathogenicity, and the amount and type of substrate. To this might be added the status of the neonate's cellular and humoral immune system, elements in the NEC equation which are being investigated today at places like SUNY Downstate Medical Center at Brooklyn. These variations may explain not only the relatively low incidence of the disease, but also its occasional fulminating course, especially in very-low-birth-weight babies, its occurrence in infants who have been fed exclusively on breast milk,[7] and its occasional appearance in term and older infants and in infants who have never been fed.[45]

Management

Following the recognition of NEC as a clinical entity in 1964, initial efforts at improvement in outcome centered mainly on medical management of the disease with emphasis on refining the indications for operation.[11,13,46] The staging system (suspect, diagnosed, advanced) recommended by Bell and his colleagues[47] has proven helpful in both guiding therapy and evaluating results.

The poor survival rate of 22% for infants with diagnosed NEC over a 20 year period at the Babies Hospital[13] led to the development of a treatment protocol for suspect and diagnosed cases without signs of necrotic bowel or perforation. This protocol, which has been followed since 1972 includes nasogastric decompression, intravenous fluid resuscitation, intravenous alimentation, parenteral antibiotics, and frequent, careful physical examination. Abdominal roentgenograms are obtained every 8 hours along with daily studies of the blood cell and platelet counts, serum electrolytes, acid-base status, and where indicated, the coagulation profile. Enteral antibiotics and intravenous low molecular weight Dextran are no longer used routinely.

In 1981 a 10-year review of diagnosed infants managed under this protocol showed an improvement in survival to 68%.[48] This improvement was almost certainly influenced by concomitant progress in neonatal critical care. However, when compared with overall infant survival, the abrupt rise in NEC survival, seen in all birth weight categories, strongly suggested that the protocol was a contributing factor. It was also noted in this study that two-thirds of the infants dying of NEC did so within 24 hours of the onset of symptoms.

Refining the indications for operation became important soon after the initial report of Touloukian[12] in 1967. Pneumatosis, in his early experience, was quickly abandoned as a sole indication for celiotomy. The extent of injury to the inner layers of the bowel could not be adequately assessed, and intestine considered to be grossly normal at the operating table progressed to necrosis and perforation.

In addition to pneumoperitoneum, the gradual recognition of other x-ray signs of necrotic bowel with or without perforation included a progressive pattern of obstruction, fixed and unchanging collections of intra- or extraluminal gas, usually in the right lower quadrant, and the sudden appearance of intraperitoneal fluid.[49]

Physical signs that were recognized to be important indicators for operation included abdominal wall erythema, induration and a palpable abdominal mass. In addition, progressive deterioration with diminished signs of peripheral perfusion, decreasing urinary output and an intractable metabolic acidosis with falling serum sodium, coagulation factors, and depressed total neutrophil count with mostly immature forms were considered further evidence of gangrene or perforation. To these criteria were added the results of selective paracentesis as advocated by Kosloske and Lilly in 1978.[50]

During the early experience with NEC, operations were primarily directed at resecting all necrotic and suspect bowel with some form of proximal ostomy. Usually the proximal and distal ends of bowel were brought out through separate stab wounds or at opposite corners of the incision. Rarely were primary anastomoses performed and only under favorable conditions in stable infants. It soon became obvious that in many instances severe short bowel resulted, especially if long segments of suspect bowel were excised. From this an early interest in second- and even third- look celiotomies at 24 and 48 hours developed in the expectation that portions of these suspect segments left in situ would be found to be salvageable.

Further surgical improvisations, including simple peritoneal drainage, high diverting ostomy, primary anastomosis, "patch and drain", and most recently "clip and drop back" methods are discussed elsewhere in this monograph.

Summary

An attempt has been made to present an overview of the early investigations into NEC. Constraints of space have made it impossible to include the many clinicians and researchers involved in the study of this fascinating and devastating disease. No separate discussion has been undertaken on the subject of prevention, although facets have been touched upon in the pages on pathogenesis. Chief among these have been the attention paid to the type and rate of enteral feedings, the carefully considered use of umbilical catheters and exchange transfusions, good maternal and perinatal care, prevention or prompt treatment of amnionitis, and stringent isolation techniques when the disease appears in the nursery. Future studies will probably focus on deficiencies and values of additives to enteral feedings, as has already been demonstrated with IgA and IgG supplements.[15] The value of breast milk is again being emphasized.

If the mortality from NEC is to be further reduced it can only be through prevention. In the author's opinion, it is unlikely that treatment will save those infants whose NEC is massive, fulminating, and leading to death within 24 hours of the onset of symptoms.[48] It is these infants who now comprise the major mortality. Furthermore, as overall mortality among low-birth-weight infants continues to decline and the smaller newborns survive, mortality from NEC may begin to increase.[52,53]

References

1. Generisch A. Bauchfellentzündung beim neugeborenen in folge von perforation des ileums. Virchows Arch Path Anat 1891; 126:485.
2. Rowe MI. Necrotizing Enterocolitis. In: Welch KJ, Randolph JG, Ravitch MM, O'Neill JA, Rowe MI, eds. Pediatric Surgery. Chicago: Year-Book Medical Publishers, Inc., 1986: 944-958.
3. Paltauf A. Die spontane dickdarm ruptur der neugeborenen. Virchows Arch Path Anat 1888; 111:461.
4. Thelander HE. Perforation of the gastrointestinal tract of the newborn infant. Am J Dis Child 1939; 58:371-393.
5. Santulli TV. personal communication.
6. Schmid KO. Über eine besonders schwer verlaufende form von enteritis beim säugling, "enterocolitis ulcerosa necroticans". I. Pathologisch-anatomische Studien. Österr Z Kinderh und Kinderf 1953; 8:114-136.
7. Quaiser K. Über eine besonders schwer verlaufende form von enteritis beim säugling, "enterocolitis ulcerosa necroticans". II. Klinische Studien. Österr Z Kinderh und Kinderf 1953; 8:136-152.
8. Waldhausen JA, Herendeen T, King H. Necrotizing colitis of the newborn. Surgery 1963; 54:365-372.
9. Berdon WE, Grossman H, Baker DH et al. Necrotizing enterocolitis in the premature infant. Radiology 1964; 83:879-887.
10. Mizrahi A, Barlow O, Berdon WE et al. Necrotizing enterocolitis in premature infants. J Pediatr 1965; 66:697-706.
11. Wilson SE, Woolley MM. Primary necrotizing enterocolitis in infants. Arch Surg 1969; 99:563-566.
12. Stevenson JK, Graham CB, Oliver TK Jr et al. Neonatal necrotizing enterocolitis. A report of twenty one cases with fourteen survivors. Am J Surg 1969; 118:260-272.
13. Santulli TV, Schullinger JN, Heird WC et al. Acute necrotizing enterocolitis in infancy: A review of 64 cases. Pediatrics 1975; 55:376-387.
14. Touloukian RJ, Berdon WE, Amoury RA et al. Surgical experience with necrotizing enterocolitis in the infant. J Pediatr Surg 1967; 2:389-401.
15. Scholander PF. The master switch of life. Sci Amer 1963; 209:92-106.
16. Johansen K. Regional distribution of circulating blood during submersion asphyxia in the duck. Acta Physiol Scand 1964; 62:1-9.
17. Corday E, Irving DW, Gold H et al. Mesenteric vascular insufficiency. Am J Med 1962; 33:365-376.
18. Elsner R, Kenney DW, Burgess K. Diving bradycardia in the trained dolphin. Nature (Lond.) 1966; 212:407-408.
19. Loyd JR. The etiology of gastrointestinal perforations in the newborn. J Pediatr Surg 1969; 4:77-84.
20. Touloukian RJ, Posch JN, Spencer R. The pathogenesis of ischemic gastroenterocolitis of the neonate: Selective gut mucosal ischemia in asphyxiated neonatal piglets. J Pediatr Surg 1972; 7:194-205.
21. Barlow B, Santulli TV, Heird WC et al. An experimental study of acute neonatal enterocolitis—the importance of breast milk. J Pediatr Surg 1974; 9:587-595.
22. Pitt J, Barlow B, Heird WC et al. Macrophages and the protective action of breast milk in necrotizing enterocolitis. Pediatr Res 1977; 11:906-909.
23. Hermann RE. Perforation of the colon from necrotizing colitis in the newborn: report of a survival and a new etiologic concept. Surgery 1965; 58:436-441.
24. Corkery JJ, Dubowitz V, Lister J et al. Colonic perforation after exchange transfusion. Br Med J 1968; 4:345-349.
25. Castor WR. Spontaneous perforation of the bowel in the newborn following exchange transfusion. Can Med Ass J 1968; 99:934-939.
26. Touloukian RJ, Kadar A, Spencer RP. The gastrointestinal complications of neonatal umbilical venous exchange transfusion: A clinical and experimental study. Pediatrics 1973; 51:36-43.

27. Rand T, Weninger M, Kolhauser C et al. Effects of umbilical artery catheterization on mesenteric hemodynamics. Pediatr Radiol 1996; 26:435-438.
28. Leake RD, Thanopoulos B, Nieberg R. Hyperviscosity syndrome associated with necrotizing enterocolitis. Am J Dis Child 1975; 129:1192-1194.
29. Hakanson DO, Oh W. Necrotizing enterocolitis and hyperviscosity in the newborn infant. J Pediatr 1977; 90:458-461.
30. Le Blanc MH, D'Cruz C, Pate K. Necrotizing enterocolitis can be caused by polycythemic hyperviscosity in the newborn dog. J Pediatr 1984; 105:804-809.
31. Book LS, Herbert JJ, Atherton SO et al. Necrotizing enterocolitis in low-birth-weight infants fed on elemental formula. J Pediat 1975; 87:602-605.
32. de Lemos RA, Rogers JH, Jr, McLaughlin W. Experimental production of necrotizing enterocolitis in newborn goats. Pediat Res 1974; 8:380.
33. Willis DM, Chabot J, Radde IC et al. Unsuspected hyperosmolality of oral solutions contributing to necrotizing enterocolitis in very-low-birth-weight infants. Pediatrics 1977; 60:535-538.
34. Polin RA, Pollack PF, Barlow B et al. Necrotizing enterocolitis in term infants. J Pediatr 1976; 89:460-462.
35. Kliegman RM. Neonatal necrotizing enterocolitis: Implications for an infectious disease. Pediatr Clin No Am 1979; 26:327-344.
36. Lawrence G, Bates J, Gaul A. Pathogenesis of neonatal enterocolitis. Lancet 1982; 1:137-139.
37. Covert RF, Barman N, Domanico RS et al. Prior enteral nutrition with human milk protects against intestinal perforation in infants who develop necrotizing enterocolitis. Pediatr Res 1995: 37:305.
38. Lucas A, Cole TJ. Breast milk and neonatal necrotizing enterocolitis. Lancet 1990; 336:1519-1523.
39. Brown EG, Sweet AJ. Preventing necrotizing enterocolitis in neonates. JAMA 1978; 240:2452-2454.
40. Goldman HI. Feeding and necrotizing enterocolitis. Am J Dis Child 1980; 134:553-555.
41. Book LS, Herbst JJ, Jung AL. Comparison of fast-and-slow-feeding rate schedules to the development of necrotizing enterocolitis. J Pediatr 1976; 89:463-466.
42. Blanc WA. Amniotic infection syndrome: Pathogenesis, morphology, and significance in circumnatal mortality. Clin Obstet Gynecol 1959; 2:705-734.
43. Hsueh H, Caplan MS, Sun X et al. Pathogenetic lesions from experimental models of necrotizing enterocolitis. Pediatr Surg Int 1992; 7:415-420.
44. Kosloske A. Pathogenesis and prevention of necrotizing enterocolitis: A hypothesis based on personal observation and a review of the literature. Pediatrics 1984; 74:1086-1092.
45. Marchildon MB, Buck BE, Abdenour G. Necrotizing enterocolitis in the unfed infant. J Pediatr Surg 1982; 17:620-624.
46. Stevenson JK, Oliver TK Jr, Graham CB et al. Aggressive treatment of neonatal necrotizing enterocolitis: 38 patients with 25 survivors. J Pediatr Surg 1971; 6:28-35.
47. Bell MJ, Ternberg JL, Feigin RD. Neonatal necrotizing enterocolitis. Therapeutic decisions based upon clinical staging. Ann Surg 1978; 187:1-7.
48. Schullinger JN, Mollitt DL, Vincur CD et al. Neonatal necrotizing enterocolitis. Survival, management, and complications: A 25 year study. Am J Dis Child 1981; 135:612-614.
49. Leonidas JC, Krasna IH, Fox HA et al. Peritoneal fluid in necrotizing enterocolitis: A radiologic sign of clinical deterioration. J Pediatr 1973; 82:672-675.
50. Kosloske AM, Lilly JR. Paracentesis and lavage for diagnosis of intestinal gangrene in neonatal enterocolitis. J Pediatr Surg 1978; 13:315-320.
51. Eibl MM, Wolf HM, Furnkranz H et al. Prevention of necrotizing enterocolitis in low-birth-weight infants by IgA-IgG feeding. N Engl J Med 1988; 319:1-7.
52. Snyder CL, Gittes GK, Murphy JP et al. Survival after necrotizing enterocolitis in infants weighing less than 100 g: 25 years' experience at a single institution. J Pediatr Surg 1997; 32:434-437.
53. Holman RC, Stoll BJ, Clarke MJ et al. The epidemiology of necrotizing enterocolitis infant mortality in the United States. Am J Pub Health 1997; 87:2026-2031.

CHAPTER 2

The Bacteriology of Necrotizing Enterocolitis

Benjamin Z. Cooper and Richard J. Scriven

As the mystery of necrotizing enterocolitis (NEC) has slowly been deciphered, and characterized in the past 33 years, it has become clear that crucial to the evolution of this disease is the bacteriologic environment in which the enteric process develops. Through this understanding, preventative measures and treatments have been developed and tested, with a certain degree of success.

NEC is an inflammatory process which develops in the gastrointestinal tract of a neonate. It is clear that there is a causal relationship between the presence of bacteria in the small intestine and the availability of metabolic substrate to support bacterial growth.[1]

Theories of Pathogenesis of NEC as Related to Bacteriology

While necessary, colonization is not sufficient for the process to develop. Once the enteric organisms have become established, disruption of mucosa and bacterial translocation become a critical aspect of the pathogenesis of this disease. Although existing research has identified infection as a definite component of NEC, it remains unclear whether a systemic inflammatory response triggered by a perinatal physiologic stress results in intestinal ischemia, bacterial overgrowth, breakdown of the mucosal barrier and bacterial translocation. It is possible that a primary infectious process results in enteroinvasion, release of inflammatory mediators, toxin production, ileus, gas production, bowel distension, increased intraluminal pressure and vascular compromise causing loss of mucosal integrity, and bacterial translocation, resulting in a systemic inflammatory response. Regardless of which entity precedes the other, it is clear that fundamental to the ongoing and escalating circular process that, ultimately results in NEC, there is an enteric infectious process.

Evidence of an Inflammatory Process in NEC

There is strong evidence that supports the concept that an infectious process is fundamental to the development of NEC. Evidence of both acute and chronic inflammation can ordinarily be found at the disease site.[1] Since 1974, there have been reports of epidemics of NEC that suggest that an infectious agent is involved in the pathogenesis of this disease. Many NEC outbreaks within a NICU are due to a common pathogen.[1-6] The absence of NEC within the sterile environment of an unborn neonate suggests the need for enteric colonization. A correlation between preventive measures that decrease the pathogenic potential of enteric organisms and a decreased incidence of disease exists.[7] Organisms on histopathologic specimens of affected bowel, pathogens from blood and other normally sterile body sites, and bacterially elaborated hydrogen gas within the bowel wall are often found in infants diagnosed with NEC. Peritoneal fluids obtained from infants with NEC during surgery often contain many neutrophilic leukocytes, suggesting infection.[1]

Necrotizing Enterocolitis, edited by Brian F. Gilchrist. ©2000 Eurekah.com.

Mechanisms of Bacterial Pathogenicity

Several mechanisms may contribute to the development of a systemic inflammatory response such as NEC. Scheifele[5] reported a strong association between a delta-like toxin produced by coagulase-negative *Staphylococcus* and NEC in 1987. *Clostridium difficile*, and methicillin-resistant *Staphylococcus aureus* also produce an enterotoxin that has been associated with NEC.[3,5,9,10] Endotoxin, a lipopolysaccharide of gram negative organisms including *Escherichia coli*, *Klebsiella*, *Enterobacter*, and *Clostridium*, has been implicated as a cause of NEC as well.[11] Endotoxin release results in septic shock by causing increased vascular permeability, disseminated intravascular coagulation, metabolic acidosis, hypotension and decreased left ventricular function.[12] Production of cytokines such as TNF, released by endotoxin-stimulated macrophages, has also been found to result in septic shock that contributes to the development of NEC. Finally, pneumatosis intestinalis has been attributed to enteric bacteria, such as *Klebsiella*, that produce gas though carbohydrate fermentation, resulting in decreased intestinal pH, increased epithelial permeability and transepithelial movement of protein which stimulates the release of inflammatory mediators from bowel mucosa.[1,13,14] These mediators may cause local vasoconstriction, resulting in intestinal ischemia. As feeding is increased, along with the substrate load to the bowel, gas production from the above mechanism results in bowel distension, which compromises mesenteric blood flow and consequently results in intestinal ischemia.[12] While many outbreaks of NEC have identified common pathogens among those affected, most have not. It is possible that there remains an unidentified pathogen that is responsible for the development of NEC[10] (Table 2.1).

Although the pathogenesis is complex, and still not yet completely defined, it is clear that central to the cascade of events that culminate in the systemic inflammatory response of NEC is a localized enteric infectious process.

Bacteriology of NEC

The healthy newborn's GI tract is sterile at birth. The colonization process occurs quickly after delivery, handling, and feeding have begun. The premature neonate, whose GI tract is at risk for developing NEC, is colonized by a group of organisms that are much different from those above. Bell[15] studied the fecal and gastric microflora of critically ill neonates and found this group of organisms to be much different than the respective group of flora in normal full-term infants. Prolonged hospital stay with increased handling, mechanical ventilation, intravenous lines, and other indwelling catheters, as well as immune deficiencies of prematurity, different feeding habits, and the frequent use of antibiotics, are all forces that contribute toward this unique group of organisms.[5,16] So, it is therefore possible that the initial colonization of the neonatal bowel is with an organism or a group of organisms that are in some way pathogenic to the preemies. After this colonization has been established and the mechanisms as described above activated, the process of NEC may ensue.

Many enteric flora of pre-term infants in an intensive care setting have been found to be associated with NEC. Organisms that grow in culture from any one of various sites may not be pathogenic. Rather they may simply represent bowel colonization or transient bacteremia in the case of positive blood cultures. Such organisms include gram-positive cocci such as coagulase-positive and negative *Staphylococci* and *Enterococci*. *Staphylococcus epidermis* was implicated by Fabia[17,18] and later confirmed by many investigators as an infectious agent in the pathogenesis of NEC. At times it was found to be the only pathogen

Table 2.1. Pathogenic mechanisms and associated organisms of NEC

Mechanism	Associated bacteria	Effect
Exotoxin	1. *Staphylococcus epidermis* 2. *Staphylococcus aureus* 3. *Escherichia coli* 4. *Clostridia*	1. Cell wall lysis 2. Tissue necrosis 3. Intestinal ischemia
Enteropathogenic	1. *Escherichia coli* 2. *Salmonella* 3. *Clostridia*	1. Invasion through bowel wall 2. Bacteremia
Gas Forming	1. *Klebsiella* 2. *Clostridia*	1. Pneumatosis intestinalis 2. Intestinal ischemia 3. Increased epithelial permeability
Endotoxin	1. *Escherichia coli* 2. *Salmonella* 3. *Klebsiella* 4. *Enterobacter* 5. *Clostridia*	1. Fever 2. Hypotension 3. Vasodilatation 4. Increased vascular permeability 5. Decreased peripheral vascular resistance 6. Disseminated intravascular coagulation 7. Cytokine production

implicated, and other times was found to be one of many organisms acting in the inflammatory process. Speer[2] first reported an association between a nonendemic mucoid strain of *Escherichia coli* and NEC, and Frantz[13] reported a similar association with *Klebsiella*. Other gram-negative rods that were later found to be associated with NEC include *Enterobacter, Pseudomonas,* and *Salmonella*.[16,19] Engel,[19] in 1974, first reported an association between the anaerobic pathogen *Clostridia* and NEC when three affected infants with NEC were found to have positive blood cultures for *Clostridia*. Association between NEC and other anaerobes include *Clostridium butyricum, Clostridium perfringens, Clostridium difficile,* and *Bacteroides*.[19]

Bacteria are the most typical organisms involved in NEC, but yeast and viruses have been implicated as offending organisms. An association with *Candida* and NEC has been reported.[16] Rotbart[9] in 1983 reported an outbreak of *Rotavirus* associated NEC. In this outbreak, a comparison of typical risk factors for NEC between affected and unaffected infants revealed no significant differences. *Rotavirus* infection was the only finding that was highly correlated with the epidemic. Other viruses that have been implicated as offending organisms of NEC include *Coronavirus, Enterovirus,* and *Coxsackie virus*.[19,20]

Some organisms found to be associated with NEC are pathogens actively involved in the enteric infection and focal to the development of NEC. Many though have colonized the bowel of the affected infant and are not responsible for an inflammatory response. Among the pathogens implicated are organisms whose virulence has been attributed to their enteroinvasive potential such as *Escherichia coli, Salmonella, Clostridium difficile* and *Clostridium butyricum*. Organisms which exert their pathogenic effect by possessing endotoxins in their cell wall include *Escherichia coli, Salmonella, Klebsiella,* and *Enterobacter*.

Table 2.2. Organisms associated with NEC

Gram negative	Gram positive	
Escherecia coli	Staphylococcus epidermis	
Enterobacter	Staphylococcus aureus	
Klebsiella	Enterococcus	
Pseudomonas		
Salmonella		
Serratia marcescens		
Anaerobic	**Yeast**	**Virus**
Clostridia butyricum	Candida	Coronavirus
Clostridia perferingens		Rotavirus
Clostridia difficle		Enterovirus
Bacteroides fragilis		Coxsackie virus

Organisms that exert their pathogenic effect by the release of exotoxin include coagulase-negative *Staphylococci* that produce delta-hemolysin, *Staphylococcus aureus, Escherichia coli*, and members of the *Clostridium* species. Some organisms that have been implicated as pathogens of NEC are not restricted to the affected infant but are among the enteric flora of the healthy newborn as well (Table 2.2).

While many investigators have found an association between certain organisms and NEC, Gupta et al reported finding a, "lack of association" between a specific infectious agent and NEC.[19] Gupta[19] suggested that the combination of infectious agents present influencing microbial adherence, toxin production and carbohydrate metabolism has greater importance in the pathogenesis of NEC compared to the specific organism present.

Controlling the Infectious Process: A Means of Prevention and Treatment of NEC

The complete pathophysiology of NEC remains obscure. Since the disease has been characterized, there have been measures taken, especially with those at risk, to decrease its incidence and improve its outcome.

Induction of intestinal maturation with prenatal or neonatal steroids has been shown to decrease the incidence of NEC.[7] Implementation of strict infection-control measures to prevent fecal and oral spread has substantially decreased the incidence of NEC after outbreaks.[7,10,21] Book[21] reported a significantly lower incidence of NEC after the implementation of infection-control measures that were designed to interrupt fecal-oral spread of an "unknown" agent. Several small trials of oral antibiotic prophylaxis have been reported, with a protective effect shown in most of them.[20] Enthusiasm for this approach has waned because of concern for development of resistant organisms.[6,10,22] Another drawback is that it interferes with normal bacterial colonization that is necessary for proper mucosal and immunologic development and for efficient caloric absorption.[10]

Although is it has been known that NEC does not occur in unfed infants, withholding enteral feeds from those infants at risk may only delay the onset of the disease at the added expense of mucosal atrophy with depressed enzyme activity and increased mucosal permeability. Enteral feeds provide intraluminal substrate promoting structural and functional maturation of the gut and acquisition of normal gut flora.[7,23] Rather than delaying enteral feeds, the modification of enteral feeds may offer some protective benefit. Decreasing the volume of feeds may decrease distension and improve mesenteric blood flow. Also, decreasing feeding volumes may improve lactose absorption, preventing its fermentation and production of hydrogen gas and avoiding the deleterious effects described above.[12] Initiating small volume feeds with steady increments, and decreasing the osmolarity of enteral feeds may prove to offer some protective benefit.[7]

It has been suggested that human milk, rich in immunoprotective factors such as secretory IgA and phagocytic cells,[1,19,24] as well as having many other potentially protective components that have not yet been clearly defined, offers protection from NEC. Administration of oral IgA-IgG immunoglobulin by bolstering intestinal immunity, has been shown to decrease the incidence of NEC.[7,8,10,12] Finally, acidification of oral fluids has been shown to reduce the incidence of NEC.[7]

It is clear that if the pathogenic means of the enteric organisms, which are at least partly responsible for this disease process, could be eradicated or blocked the disease's incidence could be decreased. Many treatment modalities have been attempted on infants diagnosed with this disease, some of them with success.

Although the role that infection plays in the development of NEC has still not been clearly defined, it is clear that NEC must be treated as an infectious disease. Empiric antibiotic administration to infants suspected of having NEC should be based on organisms prevalent in a particular nursery, especially coliforms such as *Escherichia coli*, coagulase-negative staphylococci, and enterococci. Empiric treatment of anaerobes remains controversial.[10]

Conclusion

Although many of its mysteries have been solved, much needs to be done. Investigative efforts must continue to refine those treatment options that have clinical benefit. There may be a selective role for the use of short-term prophylactic antibiotics to stop nursery outbreaks.[7] New antibiotics may prove to be more effective and less toxic than those available now. The use of oral immunoglobulin and modification of enteral feeds may ultimately prove to be more effective than the current knowledge indicates. Those individuals affected by this devastating disease can find some rest in the hope that novel and unique approaches such as competitive blockade of *Escherichia coli* transcytosis by gram-positive organisms may result in new cures.[22] Finally, efforts made to decrease the number of premature births in our society, largely through social venues, may be the most effective means of reducing the ravages of this disease.

References

1. Buescher ES. Host defense mechanisms of human milk and their relations to enteric infections and necrotizing enterocolitis. Clin Perinat 1994; 21(2):247-262.
2. Speer ME, Taber LH, Yow MD et al. Fulminant neonatal sepsis and necrotizing enterocolitis associated with a "nonenteropathogenic" strain of *Escherichia coli*. J Pediatr 1976; 89(1):91-95.
3. Overturf GD, Sherman MP, Scheifele DW et al. Neonatal necrotizing enterocolitis associated with delta toxin-producing methicillin-resistant staphylococcus aureus. Pediatr Infect Dis J 1990; 9:88-91.
4. Scheifele DW. Role of bacterial toxins in neonatal necrotizing enterocolitis. J Pediatr 1990; 117:S44-S46.
5. Scheifele DW, Bjornson GL, Dyer RA et al. Delta like toxin produced by coagulase-negative staphylococci is associated with neonatal necrotizing enterocolitis. Infection and Immunity 1987; 55(9):2268-2273.
6. Rotbart HA, Levin MJ, Yolken RH et al. An outbreak of rotavirus-associated necrotizing enterocolitis. J Pediatr 1983; 103:454-459.

7. Vasan U, Gotoff SP. Prevention of neonatal necrotizing enterocolitis. Clin Perinatol 1994; 21(2):425-435.

8. Eibl MM, Wolf HM, Furnkranz H et al. Prevention of necrotizing enterocolitis in low-birthweight infants by IgA-IgG feeding. N Eng J Med 1988; 319:1-7.

9. Rotbart HA, Johnson ZT, Reller B. Analysis of enteric coagulase-negative staphylococci from neonates with necrotizing enterocolitis. Pediatr Infect Dis J 1989; 8:140-142.

10. Willoughby RE, Pickering LK. Necrotizing enterocolitis and infection. Clin Perinatol 1994; 21(2);307-315.

11. Duffy LC, Zielezny MA, Carrion V et al. Concordance of Bacterila cultures with endotoxin and interleukin-6 in necrotizing enterocolitis. Dige Dis Scien 1997; 42(2):359-365.

12. Deitch EA. Role of bacterial translocation in necrotizing enterocolitis. Acta Paediatr 1994; Suppl 396:33-36.

13. Frantz ID, Heureux P, Engel RR et al. Necrotizing enterocolitis. J Pediatr 1975; 86(2):259-263.

14. Clark DA, Miller MJ. Intraluminal pathogenesis of necrotizing enterocolitis. J Pediatr 1990; 117:S64-S68.

15. Bell MJ, Rudinsky M, Brotherton T et al. Gastrointestinal microecology in the critically ill neonate. J Pediatr Surg 1984; 19(6):745-750.

16. Mollitt DL, Tepas JJ, Talbert JL. The microbiology of neonatal peritonitis. Arch Surg 1988; 123:176-179.

17. Mollitt DL, Tepas JJ, Talbert JL. The role of coagulase-negative staphylococcus in neonatal necrotizing enterocolitis. J Pediatr Surg 1988; 23(1):60-63.

18. Gruskay JA, Abbasi S, Anday E et al. Staphylococcus epidermis-associated enterocolitis. J Pediatr 1986; 109:520-524.

19. Gupta S, Morris JG, Panigrahi P et al. Endemic necrotizing enterocolitis: Lack of association with a specific infectious agent. Pediatr Infect Dis J 1994; 13:728-734.

20. Kosloske AM, Ulrich JA. A bacteriologic basis for the clinical presentation of necrotizing enterocolitis. J Pediatr Surg 1980; 15 (4):558-564.

21. Book LS, Overall JC, Herbst JJ et al. Clustering of necrotizing enterocolitis, interruption by infection control measures. N Eng J Med 1977; 297(18):984-986.

22. Panigrahi P, Penelope B, Karoly H et al. *Escherichia coli* transcytosis in a caco-2 cell model: Implications in neonatal enterocolitis. Pediatr Res 1996; 40:415-421.

23. Green AA, Lucas A, Lawson GR et al. Gut hormones and regulatory peptides in relation to enteral feeding, gastroenteritis, and necrotizing enterocolitis in infancy. J Pediatr 1990; 117:S24-S32.

24. Lucas A, Cole TJ. Breast milk and neonatal necrotizing enterocolitis. Lancet 1990; 336:1519-1523.

CHAPTER 3

The Immunology of Necrotizing Enterocolitis

Steven Piecuch

The pathogenesis of necrotizing enterocolitis remains incompletely understood despite considerable research efforts. In some patients, an intense exposure to a single specific risk factor appears to be temporally linked to the onset of necrotizing enterocolitis. For example, necrotizing enterocolitis in infants following exchange transfusion or birth asphyxia strongly suggests impaired blood flow or oxygen delivery as a pathogenic mechanism. A necrotizing enterocolitis-like illness may be caused by enteric infection with Clostridium perfringens, implying that some cases may be due to infection. However, the two factors most consistently associated with the development of necrotizing enterocolitis are prematurity and feeding. Approximately 90% of infants who develop necrotizing enterocolitis are premature infants who have been fed. A better understanding of how prematurity and feeding put the neonate at increased risk of developing necrotizing enterocolitis would assist in the designing strategies to prevent and to better treat this potentially devastating illness.

Although necrotizing enterocolitis progresses to systemic involvement, with hypotension and positive blood cultures and multiorgan failure in severe cases, the disease process itself begins at the level of the intestinal wall. Gastrointestinal function, including mucosal defenses and repair mechanisms, are underdeveloped in the newborn infant, particularly the premature, which may increase the risk of developing necrotizing enterocolitis. A better understanding of intestinal mucosal defenses and how they are impaired will lead to important insights into the pathogenesis of necrotizing enterocolitis.

Gastrointestinal Mucosal Immune System

The intestinal lumen is a site of exposure to microorganisms, toxins and foreign antigens. The intestinal mucosa provides an immunologic as well as a mechanical barrier preventing local gastrointestinal injury as well as systemic injury. The function of the intestinal mucosal immune system is to identify potentially harmful foreign antigens and prevent them from adhering to and damaging the intestinal mucosa and prevent them from being absorbed and causing systemic injury. While an appropriate immune response is important for host protection, inflammatory mediators such as cytokines or nitric oxide may potentially cause local or systemic injury. In addition, the development of immune tolerance to certain antigens, such as those normally present in foods, is important in order to avoid the development of local gastrointestinal or systemic allergy.[1] A problem in any of these areas could lead to mucosal injury and contribute to the development of necrotizing enterocolitis.

Necrotizing Enterocolitis, edited by Brian F. Gilchrist. ©2000 Eurekah.com.

The Peyer's patches are located predominantly in the terminal ileum and have a central role in mucosal immune function.[2-4] Each Peyer's patch contains a number of lymphoid follicles which are located underneath the muscularis mucosa. The intestinal epithelium overlying the Peyer's patch bulges into the intestinal lumen and is referred to as the dome epithelium. The dome epithelium contains specialized M (microfold) cells which take up microorganisms, particles and foreign antigens by pinocytosis and transport them across the epithelial barrier in vesicles for delivery to macrophages, dendritic cells and lymphocytes.[2-7] M cells lack an overlying mucin coat and have microfolds on their luminal surface, facilitating their uptake of foreign particles in the intestinal lumen.[8] M cells also possess specific surface receptors which promote the adherence and uptake of specific antigens. M cells may express class II major histocompatability molecules on their surface and have a role as antigen-presenting cells.

The germinal centers of the lymphoid follicles contain predominantly B lymphocytes. As a result of activation by antigen, IgA-committed B lymphocytes, as well as some activated T lymphocytes, migrate out of the germinal center and are carried by the lymphatic circulation to the mesenteric lymph nodes and from there to the systemic circulation. These immunologically active cells then migrate, in a process known as homing, to mucosal locations, including the mucosal surface of the intestine.[9] Terminal differentiation of the B lymphocytes into IgA-producing plasma cells under the influence of T lymphocytes occurs in the lamina propria. The interfollicular zones of the Peyer's patches contain predominantly T lymphocytes, the majority of which are CD4+ helper cells. These T cells have a role in assisting germinal matrix B lymphocytes in the process of switching from immature, IgM-bearing cells to mature IgA-committed B lymphocytes. These T lymphocytes also secrete cytokines which have a regulatory role in IgA production. The Peyer's patches have a critical role in recognizing and responding to antigens in the intestinal lumen. Although Peyer's patches are present in fetal life, antigenic stimulation is required for immunologic activation of the lymphoid follicles, a process which requires several postnatal weeks to occur.[4]

The intraepithelial cells are joined by tight junctions and form the epithelial barrier of the intestinal wall. Interferon-gamma, platelet-activating factor, histamine and other inflammatory mediators can cause an increase in intestinal permeability increasing the risk that foreign particles may penetrate the intestinal wall. This may result in direct tissue injury or infection and may also cause a potentially harmful local or systemic inflammatory response or the development of allergic phenomena. The intraepithelial cells have receptors for IgG, and it has been hypothesized that the bowel might be a site of IgG absorption in utero, particularly during the first and second trimesters, a period in gestation when relatively little transplacental transport of IgG occurs. However, amniotic fluid contains little IgG and it is not known whether significant absorption of IgG occurs via the intestine in utero;[9] the precise function of these receptors is unknown. The intraepithelial cells express major histocompatibility complex class II molecules postnatally in response to antigenic stimulation and function as antigen-presenting cells. Intraepithelial cells also secrete cytokines such as interleukin-6 and tumor necrosis factor-alpha in response to inflammation and have a role in the differentiation of the intraepithelial lymphocytes.[2]

Secretory IgA is produced by mucosal plasma cells which have differentiated from B lymphocytes. The mucosal plasma cell produces dimeric IgA which consists of two IgA molecules joined by a J-piece. The intestinal epithelial cell expresses a glycoprotein on its basolateral surface, the secretory component, which functions as a receptor for the dimeric

IgA. The dimeric IgA binds to the secretory component, is taken up by the epithelial cell and is then excreted at the apical end of the epithelial cell into the intestinal lumen. Secretory IgA is found in the mucin which forms a protective layer over the epithelial cell surface. Secretory IgA prevents foreign antigens from adhering to and damaging the intestinal epithelium and protects against infection by specific microorganisms. Unlike other immunoglobulins, secretory IgA does not bind complement or promote an inflammatory response.[10]

Intraepithelial lymphocytes are located between the epithelial cells in the basal portion of the intestinal epithelial layer. The intraepithelial lymphocytes are present in higher density in the proximal jejunum than in the colon. During fetal life, only 50% of the intraepithelial lymphocytes are either CD4$^+$ or CD8$^+$ and the remainder are both CD4$^-$ and CD8$^-$. Postnatally, 85-95% of the intraepithelial lymphocytes are CD8$^+$ suppressor/cytotoxic lymphocytes, unlike the lymphocytes of the Peyer's patches and lamina propria, which are predominately CD4$^+$ helper cells.[4,2] The intraepithelial lymphocytes may have a cytotoxic role and may also produce cytokines which may have a role in the regulation of the neighboring intraepithelial cells.[11] However, their precise function is not fully understood.

Both T and B lymphocytes are present in the lamina propria. The T lymphocytes are derived from the Peyer's patches and differentiate predominantly into CD4$^+$ helper cells. Sensitization with oral antigen induces predominantly Th2 lymphocytes in the lamina propria. Th2 cells secrete cytokines which are required for the differentiation of IgA-committed B cells into IgA-secreting plasma cells and which are less inflammatory than those secreted by Th1 cells. Th2 cells are particularly associated with allergic reactions to food and the predominance of Th2 cells may increase the neonates susceptibility to the development of food allergy. The predominance of Th2 lymphocytes may result in the neonate being deficient in Th1 cytokines such as interleukin-2, interferon-gamma and tumor necrosis factor-beta. At birth, IgA-producing plasma cells are not found in the lamina propria. They begin to appear as a result of antigenic stimulation by the end of the second week of life and predominate in the lamina propria by six months of age.

The mast cells are implicated in both IgE-mediated an non-IgE-mediated hypersensitivity reactions. Mast cells release inflammatory mediators such as cytokines, histamine and nitric oxide which participate in antibacterial defenses and which may also increase intestinal permeability. Both macrophages and dendritic cells are required by plasma cells for optimal antibody production. Macrophages are located in the epithelium and in the lamina propria where they have a phagocytic and antigen presenting function.[1] Dendritic cells have an antigen presenting function and predominate in the Peyer's patches.[10,12,13]

Defects in Systemic and Mucosal Immunity in the Newborn

Both mucosal and systemic immune function are impaired in the neonate, and this impairment is particularly marked in the premature. While the fetus as well as the newborn infant is capable of producing IgM in response to antigenic challenge, IgG is not produced prior to five to six months of age.[14] The fetus does acquire IgG transplacentally, but the bulk of this transport occurs in the third trimester.[4] Premature infants, particularly those less than 32 weeks gestational age, are IgG deficient at birth and may develop significant hypogammaglobulinemia with an increased risk of infection prior to beginning endogenous IgG production. In addition, the IgG which is acquired transplacentally reflects the mother's immune experience and may not fully protect the infant against antigens and microorganisms against which the mother lacks immunity. Cell mediated immunity is also impaired in the newborn. Neonatal T cells have impaired proliferative, helper and cytotoxic effects as well as impaired production of tumor necrosis factor and interferon-gamma. Impaired function of macrophages, neutrophils and monocytes has also been described in the term infant and is exacerbated in the premature.[13,14]

The mucosal immune system begins development by mid-gestation. In animals, in utero esophageal ligation results in a reduction in the number of intraepithelial lymphocytes and Peyer's patches, suggesting that antigens in swallowed amniotic fluid may stimulate the development of the fetal mucosal immune system.[15] Because the intrauterine environment is normally sterile, the mucosal immune system is immunologically naive at birth and requires antigenic stimulation in order to become fully functional.[9] This immunologic immaturity puts the newborn infant at increased risk of mucosal injury and increases the risk of systemic invasion by intestinal microorganisms. Postnatal exposure to antigens in the intestinal lumen leads to a progressive activation of the mucosal immune system, leading to an increase in the number and size of the Peyer's patches and an increase in serum and fecal immunoglobulins. IgA-producing plasma cells begin to appear in the intestinal mucosa by the end of the second week of life and become the predominant cell type by one month of age. The appearance of IgA producing plasma cells is delayed in the premature.[14,16] Antigen exposure increases with weaning, leading to an increase in the number of dendritic cells and further enhancing maturation of mucosal immunity. As antigen exposure increases, a downregulation of mucosal immunity also occurs, so that tolerance develops to antigens which are normally present in the gastrointestinal tract.[15]

Nonimmune Epithelial Barrier Functions

The nonimmune mediated defenses of the gastrointestinal mucosa include proteolytic enzyme activity, gastric acidity, peristalsis and the mucin layer which covers the epithelial surface.[17] Proteolytic enzyme activity is decreased in the newborn, which increases the potential for the intact absorption of ingested proteins. Gastric acid production is decreased in the neonatal period, especially in the premature infant, increasing the risk of microbial overgrowth.[4] The premature infant has decreased peristaltic activity, which increases the exposure of the intestinal epithelium to microorganisms, toxins and foreign antigens and increases the risk of mucosal injury and systemic absorption.[18] The mucin layer which covers the epithelium forms a protective barrier which prevents potentially harmful luminal contents from coming into direct contact with the epithelial cells. The mucin layer is composed of glycoprotein and contains secretory IgA.[13] The mucin present in the newborn infant has a lower molar ratio of carbohydrate to protein than in the adult, which may decrease its effectiveness.[18] The effectiveness of the mucin layer is further compromised by the newborn infant's lack of secretory IgA.

Intestinal permeability is increased in the neonate,[13] increasing the risk of epithelial injury, the development of food allergies and the systemic absorption of microorganisms, endotoxin and other foreign antigens. The absorption of ingested bovine serum albumin in newborn rabbits was found to be increased compared with that of older rabbits.[19] The newborn infant is capable of intact absorption of IgA, IgM and IgG contained in breast milk. Intestinal permeability increases as gestational age decreases. Beta-lactalbumin was found to be better absorbed in infants less than 33 weeks compared with those greater than 33 weeks and the absorption of lactulose and rhamnose was also found to be increased in the premature.[18] This gestational age-dependent increase in intestinal permeability is especially problematic in light of the premature infant's reduced ability to form antibody.

The response to immunization has been shown to be decreased in prematures compared with full term infants.[20,21] Infants born at less than 35 weeks gestation who were fed bovine serum albumin did not produce antibody while infants born at greater than 35 weeks did, implying deficiency in the ability of the premature to respond systemically to antigens absorbed by the enteral route.[22]

The Protective Effect of Breast Feeding

Human milk contains digestive enzymes, trophic factors and secretory IgA as well as other immunologically active components. Human milk is also better absorbed than formula and is less likely to cause allergic injury to the bowel. While unfortified human milk is not nutritionally adequate for the very low birth weight infant, breast milk fortifiers are commercially available which provide the additional protein and minerals required to meet the needs of the growing premature.[23]

Trophic factors present in human milk promote gastrointestinal growth and development, provide protection from injury and enhance repair mechanisms.[24] The neonate's higher gastric pH and decreased gastrointestinal proteolytic activity may decrease the degradation of these trophic factors and the increased permeability of the neonatal bowel may promote their intact absorption. Epidermal growth factor is present in human milk and may have a role in gastrointestinal tract recovery from injury. In animal studies, epidermal growth factor promoted intestinal growth, stimulated nucleic acid synthesis and enhanced the healing of stress-induced gastric ulcers. Human milk also contains transforming growth factors which have a role in intestinal healing and which may stimulate IgA production. Insulin-like growth factors present in milk stimulate gastrointestinal growth and have been found to enhance nitrogen balance and mucosal proliferation, reduce gut atrophy and decrease the absorption of endotoxin in septic rats. Growth hormone present in human milk has been found to promote gastrointestinal growth and development.

Human milk contains nucleotides which may promote intestinal growth and development. Supplementing the diet of rats with nucleotides was found to increase mucosal protein and DNA content and increase villous height and disaccharidase activity. Nucleotides may reduce protein leak and intestinal inflammation following experimental bowel injury. Nucleotides in breast milk may serve as nucleic acid precursors during periods of increased growth or recovery from bowel injury when supplies of endogenous nucleotides may be inadequate. Human milk contains polyamines which may have a role in intestinal growth and healing, as well as high concentrations of bombesin, which enhances growth and decreases bacterial translocation. Glutamine, the enterocyte's primary respiratory fuel and a precursor for nucleotide synthesis, is also present in breast milk. Glutamine has an important role in maintaining the structure and integrity of the neonatal intestine, preventing increases in intestinal permeability and reducing bacterial translocation. The intestinal uptake of glutamine is increased in stress states. Feeding premature infants a glutamine-supplemented formula enhanced feeding tolerance and reduced the incidence of infections.[25]

Breast milk contains high concentrations of secretory IgA which has specific activity against microorganisms and antigens to which the mother has been exposed.[14] Ingested secretory IgA is stable in the neonatal gastrointestinal tract and binds microorganisms and antigens, preventing their adherence to the epithelium of the gut. Secretory IgA derived from breast milk may protect against colonization with potentially pathogenic organisms. Additional components of breast milk may also have anti-infective activity. Lysozyme may act to damage the cell walls of pathogenic bacteria and lactoperoxidase may also have antimicrobial activity. Lactoferrin which is present in human milk may inhibit bacterial growth by binding iron and preventing its utilization by bacteria. Human milk contains fatty acids which have antiviral activity and GM1 ganglioside and globotiaose which bind bacterial toxins.[26] Complement components and cytokines such as interleukin-6 and tumor necrosis factor are present in breast milk as are T and B lymphocytes, macrophages, neutrophils and monocytes.[14]

Secretory IgA has been shown to have a specific protective effect against infection by enteric organisms.[27] The role of the other immunologically active components of human milk is less clear. Breast milk leukocytes may be less functional than those present in blood and the importance of lactoferrin in controlling bacterial growth in the intestine has been questioned. Human milk may have an important role in limiting inflammation within the intestinal lumen and protecting against injury to the bowel by inflammatory mediators. The primary role of breast milk lysozyme may be to provide negative feedback in inflammation rather than bacterial killing. Antioxidants in colostrum may protect against bowel injury caused by oxygen radicals.[26]

The intestine of infants fed human milk is colonized with bifidobacteria rather than with more pathogenic gram-negative organisms.[28] In a study comparing bacterial colonization in breast and formula fed infants, *Enterobacteriaceae* predominated in both groups early on. By the sixth day of life, bifidobacteria had become predominant in the breast-fed babies while *Enterobacteriaceae* continued to predominate in the infants who were formula fed. At one month of age, while bifidobacteria predominated in both groups, the density of colonization with bifidobacteria in the infants fed human milk was ten times greater than that of the formula fed infants.[29] In an animal model of necrotizing enterocolitis in which animals were subjected to hypoxia and then fed, intestinal colonization with gram-negative organisms and the incidence of necrotizing enterocolitis was less in the breast-fed animals compared with those fed formula.[30]

Although necrotizing enterocolitis does occur in infants fed human milk,[31,32] the incidence is less than in infants fed formula. In a study of the effect of feeding on the incidence of necrotizing enterocolitis, infants fed human milk alone had a lower incidence of necrotizing enterocolitis compared with those fed formula alone. Those infants fed both human milk and formula had an incidence of necrotizing enterocolitis which was greater than that seen in the group fed human milk exclusively, but less than that in the exclusively formula fed group.[33] Necrotizing enterocolitis is a common complication following surgery to repair gastroschisis, with a reported incidence as high as 20%. In a large series of infants who underwent gastroschisis repair, the incidence of necrotizing enterocolitis was found to be highest in those infants who received formula alone, of intermediate incidence in those infants fed formula and human milk and least in those infants fed human milk exclusively.[34]

The mechanism by which breast feeding protects against the development of necrotizing enterocolitis is not fully understood and may not be limited to the immunologic benefits of breast milk. The protective effects of secretory IgA and the promotion of intestinal colonization with bifidobacteria rather than with pathogenic gram negative organisms seem to be very important. The enhanced absorption of breast milk compared to formula is beneficial, especially in light of evidence that bacterial metabolism of malabsorbed carbohydrate and protein may lead to bowel injury.[9,35] The reduced allergenicity of breast milk may reduce the risk of feeding-induced mucosal injury. The trophic factors which are present in breast milk and which promote intestinal adaptation and repair may be particularly important. Regardless of the precise mechanism of protection, the benefits of feeding sick and premature infants human milk are clear.

The Role of Systemic and Enteral Immunoglobulins

Because newborn infants, especially prematures, are deficient in both serum IgG and secretory IgA, the possible protective effects of supplemental intravenous and enteral immunoglobulins on the incidence of necrotizing enterocolitis have been investigated. The prophylactic administration of intravenous immunoglobulins has not been found to reduce the incidence of necrotizing enterocolitis,[36] even when serum IgG was increased to a level considered to be protective.[37] Providing immunoglobulins by the enteral route may decrease the risk of developing necrotizing enterocolitis by providing local mucosal protection as well as by potentially reducing intestinal colonization with pathogenic organisms. Secretory IgA is known to provide specific protection against enteral infection. In a rabbit model, the enteral administration of human secretory IgA decreased bacterial translocation.[38] Feeding bovine milk immunoglobulin with high titers of antibody against enterotoxigenic *E. coli* was found to be effective in preventing diarrhea following challenge with that organism.[27] Infants who were fed an IgG preparation had a delayed onset and reduced severity of diarrhea due to rotavirus as well as a reduced incidence of necrotizing enterocolitis.[39]

One group of investigators found a significant reduction in the incidence of necrotizing enterocolitis in a group of infants fed an IgA-IgG preparation. The enterally administered immunoglobulin was not absorbed suggesting that the enterally administered immuno-globulin acted locally at the epithelial surface. It was hypothesized that the IgG had an antitoxic and opsonizing effect and that the IgA had a mucosal protective effect.[40] It is also possible that the immunoglobulin acted locally to decrease the release of potentially harmful inflammatory mediators. In the laboratory, adding an IgA-IgG preparation to monocytes was found to result in a dose-dependent decrease in the release of interleukin-6 and tumor necrosis factor-alpha, both cytokines with the ability to cause tissue damage.[41] The beneficial effect of enteral IgA-IgG on the incidence of necrotizing enterocolitis was not confirmed by another group of investigators, who found that an enteral IgA-IgG preparation was less effective in reducing the incidence of necrotizing enterocolitis than feeding infants a prepa-ration of oral gentamicin plus lyophilized enterobacteria.[42]

Beneficial Effects of Steroids in Reducing Necrotizing Enterocolitis

Administering antenatal steroids to pregnant women at risk of preterm delivery has been shown to enhance pulmonary maturation and to reduce the incidence of respiratory distress syndrome in their infants.[43-45] Antenatal steroids have also been shown to reduce the incidence of necrotizing enterocolitis in the neonate in some studies,[43,45,46] while in other studies no beneficial effect was seen.[44,47] Postnatal steroids may also reduce the incidence of necrotizing enterocolitis. Premature infants who received postnatal dexamethasone to prevent the development of bronchopulmonary dysplasia had a decreased incidence of necrotizing enterocolitis compared with an untreated control group.[48] Another study found that a group of infants who received postnatal steroids had an incidence of necrotizing enterocolitis which was less than a group who received no steroids at all and greater than a group treated with antenatal steroids, implying a beneficial effect of postnatal steroid therapy.[49] Steroids are thought to exert their protective effect by enhancing intestinal maturation. In experimental work in rodents, steroids were found to strengthen the intestinal mucosa, increase resistance to ischemia and decrease bacterial translocation.[50,51] The postnatal administration of hydrocortisone in rats was found to increase the activity of diamine oxidase, an enzyme thought to have a protective effect in ischemia.[52] Steroids may exert their beneficial effect by other mechanisms as well. For example, infants who received antenatal steroids were found to have less hypotension and less of a need for vasopressor support in the first 48 hours of life compared with a control group.[53]

Bacterial Translocation in Necrotizing Enterocolitis and the Protective Effect of Enteral Antibiotics

Bacterial translocation is the process by which bacteria and their products, such as endotoxin, cross the intestinal epithelial barrier and gain entry to the intestinal tissue and ultimately the systemic circulation. An intact epithelial barrier and a functioning mucosal immune system are both important in preventing bacterial translocation. The systemic immune system is important in controlling systemic infection after intestinal organisms have crossed the epithelial barrier. The newborn infant, especially the premature, is at high risk for bacterial translocation because of increased intestinal permeability and impaired mucosal and systemic immune function. Although bacterial colonization of the intestine is a normal process and exposure to microorganisms as well as to other luminal antigens has an important role in stimulating the development of the mucosal immune system, bacterial overgrowth, especially with gram-negative organisms, is potentially harmful. Hospitalization in a neonatal intensive care unit and the use of broad spectrum antibiotics are associated with changes in intestinal flora, bacterial overgrowth and colonization with gram-negative organisms.[54] One potential protective effect of breast feeding is the promotion of intestinal colonization with bifidobacteria rather than with gram negative bacteria.[17]

Gram-negative organisms are particularly prone to bacterial translocation.[55] In a mouse model, gram-negative organisms were found to have a higher rate of translocation than gram-positive organisms or anaerobes. The most common bacteria isolated from blood cultures of patients with necrotizing enterocolitis are gram negatives, emphasizing the important role intestinal colonization with gram-negative organisms has in the pathogenesis of necrotizing enterocolitis. Intestinal ischemia may increase intestinal permeability and increase the risk of bacterial translocation.[56] Whether gram-negative organisms exert their harmful effect by causing primary intestinal injury or by invading an already damaged mucosa is unclear. In animal models, endotoxin causes intestinal injury resembling that seen in infants with necrotizing enterocolitis.[57,58]

Evidence from animal studies suggests that bacterial overgrowth may increase intestinal permeability and that oral antibiotic therapy may reverse this effect.[59] In an animal model, ischemia in the absence of intestinal bacteria caused stricture rather than perforation, supporting a potential role for enteral as well as parenteral antibiotics in reducing the risk of perforation in infants with necrotizing enterocolitis.[60] Enteral antibiotics have been used to limit the growth of pathogenic intestinal organisms in an effort to reduce the risk of developing necrotizing enterocolitis. Oral gentamicin and oral kanamycin were found by some to decrease the density of colonization with gram-negative organisms[56] and to reduce the incidence of necrotizing enterocolitis.[61,62] Others found no significant reduction in the incidence of necrotizing enterocolitis with the use of enteral gentamicin[63] or enteral kanamycin. An increase in colonization with kanamycin-resistant organisms was observed in infants placed on routine kanamycin prophylaxis.[64] Oral vancomycin was found to be effective in terminating an outbreak of necrotizing enterocolitis associated with *Clostridium difficile*.[65]

Oral antibiotics may be useful in controlling outbreaks of necrotizing enterocolitis linked to a particular organism. In such cases, an appropriate, relatively narrow spectrum oral antibiotic can be used to reduce colonization with the offending organism and reduce the risk of necrotizing enterocolitis. However, the widespread use of oral antibiotics is less likely to be successful, since necrotizing enterocolitis has been linked to both gram-positive and gram-negative organisms, as well as viruses and yeast. Routine use of oral antibiotics may not only be ineffective but may also result in an increase in gastrointestinal colonization with resistant organisms.

Cytokines and Inflammatory Mediators

In animal experiments the injection of inflammatory mediators such as endotoxin, tumor necrosis factor and platelet-activating factor leads to hypotension and bowel necrosis resembling that seen in infants with necrotizing enterocolitis.[57,66,9,58] Platelet-activating factor, a phospholipid produced by inflammatory cells, platelets, endothelial cells and bacteria such as *E. coli*, is a potent mediator of hypotension and intestinal necrosis. Platelet-activating factor induces the formation of tumor necrosis factor as well as the production of more platelet-activating factor.[57] Administering platelet-activating factor antagonist prior to the injection of lipopolysaccharide resulted in a lower level of intestinal platelet-activating factor, suggesting that platelet-activating factor itself causes the release of additional platelet-activating factor in a positive feedback mechanism.[67]

Tumor necrosis factor, a product of monocytes and macrophages as well as of lymphocytes and other tissues, also causes hypotension and intestinal necrosis. By itself, the effects of tumor necrosis factor are relatively mild.[57] When tumor necrosis factor and platelet-activating factor are administered together, the result is profound hypotension and tissue injury. Tumor necrosis factor and interleukin-6 were found to be increased in patients with confirmed necrotizing enterocolitis and the levels of interleukin-6, but not of tumor necrosis factor, were found to correlate with the severity of the necrotizing enterocolitis.[68] The administration of either tumor necrosis factor or lipopolysaccharide alone in low dosage did not produce intestinal necrosis. When both were given together, gross and microscopic injury similar to that seen in necrotizing enterocolitis was produced. The increase in intestinal levels of platelet-activating factor when both were given together was greater than when either was given alone, and the use of platelet-activating factor antagonists prevented the intestinal injury. The administration of high dose tumor necrosis factor also caused bowel necrosis. Again, intestinal levels of platelet-activating factor were increased and the effect was blocked by platelet-activating factor antagonists.[69]

In a rat model, necrotizing enterocolitis was caused by the intraaortic injection of platelet-activating factor.[70] Subjecting rats to hypoxia caused a decrease in intestinal perfusion and an increase in platelet-activating factor level. Histologic examination of the intestine revealed early necrotic changes. Pretreatment with platelet-activating factor antagonist prevented the decrease in intestinal perfusion and prevented the necrotic changes.[71] Subjecting rats to bacterial colonization, formula feeding and asphyxia caused elevation in platelet-activating factor levels and the development of necrotizing enterocolitis. The use of antagonist resulted in a decrease in the platelet-activating factor levels and a reduction in the incidence of necrotizing enterocolitis.[72]

Lipopolysaccharide, the endotoxin of *E. coli*, induces the production and release of cytokines such as platelet-activating factor and tumor necrosis factor. The injection of lipopolysaccharide in animals causes shock and tissue necrosis, which is blocked by the administration of antagonists to platelet-activating factor and tumor necrosis factor. In a rat model, platelet-activating factor levels increased in response to hypoxia and caused intestinal tissue injury which was ameliorated by pretreatment with platelet-activating factor antagonists. The tissue injury observed was relatively mild and was not as severe as expected in clinical necrotizing enterocolitis. Adding lipopolysaccharide increased intestinal platelet-activating factor levels and increased bowel necrosis,[57] implying an important role for microbial toxins in the pathogenesis of necrotizing enterocolitis. The injection of lipopolysaccharide in an animal model caused a ten-fold increase in platelet-activating factor levels and bowel necrosis which was prevented by the use of platelet-activating factor antagonist.[67] The finding that toxins may cause intestinal injury and may cause the release of potentially harmful inflammatory mediators suggests at least one mechanism by which intestinal overgrowth or colonization with pathogenic organisms may lead to the development of necrotizing enterocolitis.

Platelet-activating factor levels have been found to be elevated in infants with necrotizing enterocolitis and the degree of elevation has been shown to correlate with the severity of the illness. Platelet-activating factor may act in part through the release of superoxide and other oxygen radicals by neutrophils. In rats, depletion of neutrophils by vinblastine resulted in a decrease in the local and systemic effects of platelet-activating factor and lipopolysaccharide toxin.[73] Pretreatment with superoxide dismutase, catalase or deferoxamine reduced the increase in intestinal permeability seen when platelet-activating factor is administered in low dosage. However, when platelet-activating factor was administered in high dosage, the administration of superoxide dismutase, catalase or deferoxamine had no effect, suggesting that while low dose platelet-activating factor may act through oxygen radicals, high dose platelet-activating factor works through other mechanisms.[74]

Levels of platelet-activating factor-acetylhydrolase, the enzyme which degrades platelet-activating factor, is low in the newborn,[57,9,75] which may increase the susceptibility of the neonate to platelet-activating factor-induced injury. In a rat model of necrotizing enterocolitis, the enteral administration of platelet-activating factor-acetylhydrolase was found to reduce the incidence of necrotizing enterocolitis.[76] Platelet-activating factor-acetylhydrolase is present in human milk, suggesting one mechanism by which breast feeding may protect against necrotizing enterocolitis.[77,78] In rats, lowering platelet-activating factor-acetylhydrolase levels pharmacologically decreased the dosage of platelet-activating factor required to produce bowel necrosis, and the use of dexamethasone to increase platelet-activating factor-acetylhydrolase levels protected against platelet-activating factor-induced bowel injury.[70] The finding that lipopolysaccharide, tumor necrosis factor and platelet-activating factor act together to cause necrotizing enterocolitis-like lesions in animals, and that platelet-activating factor antagonists are protective in animals, suggests a possible role for the clinical use of platelet-activating factor antagonists in the prevention and treatment of human necrotizing enterocolitis.

Nitric oxide is the product of the conversion of arginine to citrulline and may have an important role in the pathogenesis of necrotizing enterocolitis. Nitric oxide synthase exists in three isoforms, all of which are present in the intestine. Neuronal nitric oxide synthase (NOS-1) and endothelial nitric oxide synthase (NOS-3) are both expressed constitutively and produce relatively small amounts of nitric oxide. Inducible nitric oxide synthase (NOS-2) may be induced by inflammatory cytokines and lead to the production of relatively large amounts of nitric oxide. The constitutive production of nitric oxide may protect the intestine through its vasodilatory and antiplatelet effects. The inducible production of relatively large amounts of nitric oxide may have antimicrobial activity but may also increase intestinal permeability and cause cellular injury. For example, lipopolysaccharide causes an upregulation of inducible nitric oxide synthase activity which promotes bacterial killing but which may also cause intestinal injury.[79]

Experimental work with nitric oxide has given results which may initially appear to be inconsistent. Inhibition of nitric oxide caused an increase in intestinal permeability, which was prevented by the administration of L-arginine, a nitric oxide precursor, as well as by nitroprusside, which causes the release of nitric oxide. Nitroglycerin, a nitric oxide donor, was found to have a protective effect in a rabbit model of necrotizing enterocolitis.[80] Nitric oxide may protect against inflammatory injury by inactivating superoxide and preventing the adherence of leukocytes.[81] In rats, injection of lipopolysaccharide induced nitric oxide synthase activity in intestinal epithelial cells, which was associated with a decrease in cell viability. Dexamethasone inhibited this increase in nitric oxide synthase activity and had a protective effect.[82] Nitric oxide inhibition has also been found to reduce bacterial translocation following lipopolysaccharide challenge. Nitric oxide levels have been found to be increased in patients with inflammatory bowel disease. In a guinea pig model of inflammatory bowel disease, nitric oxide inhibition has been shown to reduce disease activity. It has been hypothesized that the constitutive production of nitric oxide has a vasodilatory effect which is important in maintaining intestinal perfusion. The sustained upregulation of nitric oxide production as occurs in inflammation leads to mucosal injury and increased permeability. A better understanding of the role of nitric oxide in the pathogenesis of necrotizing enterocolitis may lead to new strategies for prevention and treatment.[66]

Summary

Most cases of necrotizing enterocolitis are the result of multiple causative factors and a better understanding of how these causative factors act suggests preventative and therapeutic strategies. Breast feeding may protect by preventing colonization by pathogenic organisms, by providing secretory IgA and other immunologically active protective factors and by providing trophic factors which enhance gastrointestinal development and repair. Cautious feeding practices may reduce malabsorption and reduce the potential for intestinal injury from the microbial metabolism of malabsorbed carbohydrate. The potential role of steroids, enteral IgA and the platelet-activating factor warrant further investigation. The role of constitutive and inducible nitric oxide synthase in the pathogenesis of necrotizing enterocolitis is better understood.

Necrotizing enterocolitis is a disease which begins at the level of the intestinal epithelium. The impaired mucosal defenses of the newborn infant, particularly the premature, set the stage for its development. A better understanding of the mechanisms of how these defenses operate and how they are impaired in the newborn infant, particularly in the premature, will assist us in better understanding how the various risk factors implicated in the pathogenesis of necrotizing enterocolitis actually cause disease in the infant.

Necrotizing enterocolitis seems to be the result of inflammation. The newborn infant's impairment of mucosal immune function, in concert with the deficiencies in nonimmune intestinal barrier function and systemic immune function, appears to be a central factor putting the neonate, particularly the premature, at risk for necrotizing enterocolitis.

References

1. Hanson LA, Hahn-Zoric M, Wiedermann U et al. Early dietary influence on later immunocompetence. Nutr Rev 1996; 54:S23-S30.
2. McGowan I, Chalmers A, Smith G-R et al. Advances in mucosal immunology. Gastroenterol Clin North Am 1997; 26:145-173.
3. Cornes JS. Number, size, and distribution of Peyer's patches in the human small intestine. Gut 1965; 6:225-233.
4. Insoft RI, Sanderson IR, Walker WA. Development of immune function in the intestine and its role in neonatal diseases. Pediatr Clin North Am 1996; 43:551-571.
5. Wolf JL, Rubin DH, Finberg R et al. Intestinal M cells: A pathway for entry of retrovirus into the host. Science 1981; 212:471-472.
6. Sicinsski P, Rowinski J, Warchol JB et al. Poliovirus type 1 enters the human host through the intestinal M cells. Gastroenterology 1990; 98:56-58.
7. Wassef JS, Keren DF, Mailloux JL. Role of M cells in initial antigen uptake and in ulcer formation in the rabbit intestinal loop model of shigellosis. Infect Immun 1989; 57:858-863.
8. Owen RL, Pierce NF, Apple RT et al. M cell transport of Vibrio cholerae from the intestinal lumen into Peyer's patches: a mechanism for antigen sampling and for microbial transepithelial migration. J Infect Dis 1986; 153:1108-1118.
9. Kleigman RM, Walker WA, Yolken RH. Necrotizing enterocolitis: research agenda for a disease of unknown etiology and pathogenesis. Pediatr Res 1993; 34:701-708.
10. Toy LS, Mayer L. Basic and clinical overview of the mucosal immune system. Semin Gastrointest Dis 1996; 7:2-11.
11. Beagley KW, Husband AJ. Intraepithelial lymphocytes: origins, distribution, and function. Crit Rev Immunol 1998; 18:237-254.
12. Mannick E, Udall Jr JN. Neonatal gastrointestinal mucosal immunity. Clin Perinatol 1996; 23:287-304.
13. Udall Jr JN. Gastrointestinal host defense and necrotizing enterocolitis. J Pediatr 1990; 117:S33-S43.
14. Groer M, Walker WA. What is the role of preterm breast milk supplementation in the host defenses of preterm infants? Fact vs. fiction. Adv Pediatr 1996; 43:335-358.
15. Cummins AG, Thompson FM. Postnatal changes in mucosal immune response: a physiological perspective of breast feeding and weaning. Immunol Cell Biol 1997; 75:419-429.
16. Gleeson M, Cripps AW, Clancy RL. Modifiers of the mucosal immune system. Immunol Cell Biol 1995; 73:397-404.
17. Deitch EA. Role of bacterial translocation in necrotizing enterocolitis. Acta Paediatrica (Suppl) 1994; 396:33-36.
18. Israel EJ. Neonatal necrotizing enterocolitis, a disease of the immature intestinal mucosal barrier. Acta Paediatr (Suppl) 1994; 396:27-32.
19. Udall JN, Pang K, Fritze L et al. Development of gastrointestinal mucosal barrier. I. The effect of age on intestinal permeability to macromolecules. Pediatr Res 1981; 15:241-244.
20. Bernbaum JC, Daft A, Anolik R et al. Response of preterm infants to diphtheria-tetanus-pertussis immunizations. J Pediatr 1985; 107:184-188.
21. Bernbaum J, Daft A, Anolik R et al. Half-dose immunization for diphtheria, tetanus, pertussis: response of preterm infants. Pediatrics 1989; 83:471-476.
22. Rieger CHL, Rothberg RM. Development of the capacity to produce specific antibody to an ingested food antigen in the premature infant. J Pediatr 1975;87:515-518.
23. Schanler RJ. Suitability of human milk for the low-birthweight infant. Clin Perinatol 1995; 22:207-222.
24. Lo CW, Kleinman RE. Infant formula, past and future: Opportunities for improvement. Am J Clin Nutr 1996; 63:646S-650S.
25. Carver JD, Barness LA. Trophic factors for the gastrointestinal tract. Clin Perinatol 1996; 23:265-285.
26. Buescher ES. Host defense mechanisms of human milk and their relations to enteric infections and necrotizing enterocolitis. Clin Perinatol 1994; 21:247-262.

27, Tacket CO, Losonsky G, Link H et al. Protection by milk immunoglobulin concentrate against oral challenge with enterotoxigenic *Escherichia coli*. N Engl J Med 1988; 318:1240-1243.
28. Balmer SE, Wharton BA. Diet and faecal flora in the newborn: breast milk and infant formula. Arch Dis Child 1989; 64:1672-1677.
29. Yoshioka H, Iseki KI, Fujita K. Development and differences of intestinal flora in the neonatal period in breast-fed and bottle-fed infants. Pediatrics 1983; 72:317-321.
30. Barlow B, Santulli TV, Heird WC et al. An experimental study of acute neonatal enterocolitis-the importance of breast milk. J Pediatr Surg 1974; 9:587-595.
31. Kleigman RM, Pittard WB, Fanaroff AA. Necrotizing enterocolitis in neonates fed human milk. J Pediatr 1979; 95:450-453.
32. Moriartey RR, Finer NN, Cox SF et al. Necrotizing enterocolitis and human milk. J Pediatr 1979; 94:295-296.
33. Lucas A, Cole TJ. Breast milk and neonatal necrotizing enterocolitis. Lancet 1990; 336:1519-1523.
34. Jayanthi BS, Seymour P, Puntis JWL et al. Necrotizing enterocolitis after gastroschisis repair: a preventable complication? J Pediatr Surg 1998; 33:705-707.
35. DiLorenzo M, Bass J, Krantis A. Use of L-arginine in the treatment of experimental necrotizing enterocolitis. J Pediatr Surg 1995; 30:235-241.
36. Malik S, Giacoia GP, West K. The use of intravenous immunoglobulin (IVIG) to prevent infections in bronchopulmonary dysplasia: Report of a pilot study. J Perinatol 1991; 11:239-244.
37. Fanaroff AA, Korones SB, Wright L et al. A controlled trial of intravenous immune globulin to reduce nosocomial infections in very-low-birth-weight infants. N Engl J Med 1994; 330:1107-1113.
38. Maxson RT, Jackson RJ, Smith SD. The protective role of enteral IgA supplementation in neonatal gut origin sepsis. J Pediatr Surg 1995; 30:231-234.
39. Barnes GL, Doyle LW, Hewson PH et al. A randomized trial of oral gammaglobulin in low-birth-weight infants infected with rotavirus. Lancet 1982; 1:1371-1373.
40. Eibl MM, Wolf HM, Furnkranz H, Rosenkranz A. Prevention of necrotizing enterocolitis in low-birth weight infants by IgA-IgG feeding. N Engl J Med 1988; 319:1-7.
41. Wolf HM, Eibl MM. The anti-inflammatory effect of an oral immunoglobulin (IgA-IgG) preparation and its possible relevance for the prevention of necrotizing enterocolitis. Acta Paediatr (Suppl) 1994; 396:37-40.
42. Fast C, Rosegger H. Necrotizing enterocolitis prophylaxis: oral antibiotics and lyophilized enterobacteria vs. oral immunoglobulins. Acta Paediatrica (Suppl) 1994; 396:86-90.
43. Ballard RA, Ballard PL. Antenatal hormone therapy for improving the outcome of the preterm infant. J Perinatol 1996; 16:390-396.
44. Maher JE, Cliver SP, Goldenberg RL et al. The effect of corticosteroid therapy in the very premature infant. Am J Obstet Gynecol 1994; 170:869-873.
45. Crowley P. Corticosteroids after preterm premature rupture of membranes. Obstet Gynecol Clin North Am 1992; 19:317-326.
46. Bauer CR, Morrison JC, Poole WK et al. A decreased incidence of necrotizing enterocolitis after prenatal glucocorticoid therapy. Pediatrics 1984; 73:682-688.
47. Silver RK, Vyskocil C, Solomon SL et al. Randomized trial of antenatal dexamethasone in surfactant-treated infants delivered before 30 weeks gestation. Obstet Gynecol 1996; 87:683-691.
48. Tapia JL, Ramirez R, Cifuentes J et al. The effect of early dexamethasone administration on bronchopulmonary dysplasia in preterm infants with respiratory distress syndrome. J Pediatr 1998; 132:48-52.
49. Halac E, Halac J, Begue EF et al. Prenatal and postnatal corticosteroid therapy to prevent neonatal necrotizing enterocolitis: a controlled trial. J Pediatr 1990; 117:132-138.
50. Israel EJ, Schiffrin EJ, Carter EA et al. Cortisone strengthens the intestinal mucosal barrier in a rodent necrotizing enterocolitis model. Adv Exp Med Biol 1991; 310:375-380.
51. Israel EJ, Schiffrin EJ, Carter EA et al. Prevention of necrotizing enterocolitis in the rat with prenatal cortisone. Gastroenterology 1990; 99:1333-1338.
52. Karp WB, Robertson AF, Kanto WP Jr. The effect of hydrocortisone, thyroxin, and phenobarbital on diamine oxidase activity in newborn rat intestine. Pediatr Res 1987; 21:368-370.
53. Moise AA, Weardon ME, Kozinetz CA et al. Antenatal steroids are associated with less need for blood pressure support in extremely premature infants. Pediatrics 1995; 95:845-850.

54. Bell MJ, Rudinsky M, Brotherton T et al. Gastrointestinal microecology in the critically ill neonate. J Pediatr Surg 1984; 19:745-751.
55. Steffen EK, Berg RD, Deitch EA. Comparison of translocation rates of various indigenous bacteria from the gastrointestinal tract to the mesenteric lymph node. J Infect Dis 1988; 157:1032-1038.
56. Bell MJ, Shackelford PG, Feigin RD et al. Alterations in gastrointestinal microflora during antimicrobial therapy for necrotizing enterocolitis. Pediatrics 1979; 63:425-428.
57. Hsueh A, Caplan MS, Sun X et al. Platelet-activating factor, tumor necrosis factor, hypoxia, and necrotizing enterocolitis. Acta Paediatr (Suppl) 1994; 396:11-17.
58. Bhatia AM, Ramos CT, Scott SM et al. Developmental susceptibility to intestinal injury by platelet-activating factor in the newborn rat. J Invest Surg 1996; 9:351-358.
59. Rutgers HC, Batt RM, Proud FJ et al. Intestinal permeability and function in dogs with small intestinal overgrowth. J Small Animal Pract 1996; 37:428-434.
60. Bell MJ, Kosloske AM, Benton C et al. Neonatal necrotizing enterocolitis: prevention of perforation. J Pediatr Surg 1973; 8:601-605.
61. Egan EA, Mantilla G, Nelson RM et al. A prospective controlled trial of oral kanamycin in the prevention of neonatal necrotizing enterocolitis. J Pediatr 1976; 89:467-470.
62. Grylack LJ, Scanlon JW. Oral gentamicin therapy in the prevention of neonatal necrotizing enterocolitis. A controlled double-blind trial. Am J Dis of Childr 1978; 132:1192-1194.
63. Rowley MP, Dahlenburg GW. Gentamicin in the prophylaxis of neonatal narcotizing enterocolitis. Lancet 1978; 2:532.
64. Boyle R, Nelson JS, Stonestreet BS et al. Alterations in stool flora resulting from oral kanamycin prophylaxis of necrotizing enterocolitis. J Pediatr 1978; 93:857-861.
65. Han VK, Sayed H, Chance GW et al. An outbreak of Clostridium difficile necrotizing enterocolitis: a case for oral vancomycin therapy? Pediatrics 1983; 71:935-941.
66. Ford H, Watkins S, Reblock K et al. The role of inflammatory cytokines and nitric oxide in the pathogenesis of necrotizing enterocolitis. J Pediatr Surg 1997; 32:275-282.
67. Hsueh W, Gonzalez-Crussi F, Arroyave JL. Platelet-activating factor: an endogenous mediator for bowel necrosis in endotoxemia. FASEB J 1987; 1:403-405.
68. Morecroft JA, Spitz L, Hamilton PA et al. Plasma cytokine levels in necrotizing enterocolitis. Acta Paediatr (Suppl) 1994; 396:18-20.
69. Kubes P. Nitric oxide modulates epithelial permeability in the feline small intestine. Am J Physiol (Gastrointest Liver Physiol) 1992; 262:G1138-1142.
70. Furukawa M, Lee EL, Johnston JM. Platelet-activating factor-induced ischemic bowel necrosis: the effect of platelet-activating factor acetylhydrolase. Pediatr Res 1993; 34:37-41.
71. Caplan MS, Sun X-M, Hsueh W. Hypoxia causes ischemic bowel necrosis in rats: The role of platelet-activating factor (PAF-acether). Gastroenterology 1990; 99:979-986.
72. Caplan MS, Hedlund E, Adler L et al. The platelet-activating factor receptor antagonist WEB 2170 prevents neonatal necrotizing enterocolitis in rats. J Pediatr Gastroenterol Nutr 1997; 24:296-301.
73. Musemeche C, Caplan M, Hsueh W et al. Experimental necrotizing enterocolitis: the role of polymorphonuclear neutrophils. J Pediatr Surg 1991; 26:1047-1050.
74. Kubes P, Arfors KE, Granger DN. Platelet-activating factor-induced mucosal dysfunction: role of oxidants and granulocytes. Am J Physiol 1991; 260 (Gastrointest Liver Physiol 3):G965-G971.
75. Caplan M, Hsueh W, Kelly A et al. Serum PAF acetylhydrolase increases during neonatal maturation. Prostaglandins 1990; 39:705-714.
76. Caplan MS, Lickerman M, Adler L et al. The role of recombinant platelet-activating factor acetylhydrolase in a neonatal rat model of necrotizing enterocolitis. Pediatr Res 1997; 42:779-783.
77. Furukawa M, Narahara H, Yasuda K et al. Presence of platelet-activating factor-acetylhydrolase in milk. J Lipid Res 1993; 34:1603-1609.
78. Moya FR, Eguchi H, Zhao B et al. Platelet-activating factor acetylhydrolase in term and preterm human milk: a preliminary report. J Pediatr Gastroenterol Nutr 1994; 19:236-239.
79. Ford H, Watkins S, Reblock K et al. The role of inflammatory cytokines and nitric oxide in the pathogenesis of necrotizing enterocolitis. J Pediatr Surg 1997; 32:275-282.
80. Graf JL, VanderWall KJ, Adzick NS et al. Nitroglycerin attenuates the bowel damage of necrotizing enterocolitis in a rabbit model. J Pediatr Surg 1997; 32:283-285.

81. Gaboury J, Woodman RC, Granger DN et al. Nitric oxide prevents leukocyte adherence: role of superoxide. Am J Physiol 1993; 265 (Heart Circ Physiol 34):H862-H867.
82. Tepperman BL, Brown JF, Whittle BJR. Nitric oxide synthase induction and intestinal epithelial cell viability in rats. Am J Physiol 1993; 265 (Gastrointest Liver Physiol 28):G214-G218.

CHAPTER 4

The Radiology of Necrotizing Enterocolitis

E. Christine Wallace

Necrotizing enterocolitis can present insidiously. Our chief neonatologist at Tufts believes that in the management of preterm infant illness the "first chance to treat is the only chance." Thus the subtlety of presentation and delicacy of the patient demand great vigilance on the part of all. The radiologist may contribute by sustaining a high index of suspicion for features of NEC when reading all studies—both for the early sometimes nonspecific features of developing bowel disease as well as the blatant classical radiological features of established NEC.

Techniques of Evaluation

The mainstay of radiological evaluation is the supine abdominal film with supplementary cross table lateral or left lateral decubitus views. Contrast evaluation for other suspected illness may illustrate disease in the acute phase and aid assessment of complications in the recovered infant. Ultrasound evaluation has a place although the features suggesting NEC may be found serendipitously rather than by design. Similarly the small portions of the abdomen included on other studies especially the chest x-ray should be reviewed carefully as clues to developing bowel disease may be detected.

Indications

The indications for evaluation include increased aspirates or residuals from the nasogastric tube, bile stained aspirates, vomiting, constipation or diarrhea, frankly bloody stools, abdominal distension, abdominal discoloration, palpable bowel loops, lethargy, increased apneas and, or bradycardias and other signs of sepsis. Sometimes it is difficult even for an experienced neonatologist to express in words the indication for evaluation as the nuances of the bedside assessment must be incorporated into the decision.

Radiographic Projections

The supine KUB is usually the first deliberate evaluation (Fig. 4.1). A prone view may be substituted if the child is being nursed in this position. Occasionally it may be added to distinguish pneumatosis from feces mixed with air, as the colonic and rectal gas will redistribute in this position.

A horizontal beam film is the most reliable technique for the detection of free air. The cross table lateral is the preferred projection. This can be obtained without disturbing the patient and free air can be seen clearly beneath the dome of the anterior abdominal wall (Fig. 4.2). Occasionally, a left lateral decubitus view is used when doubt persists or to satisfy the prior experience of a particular clinician (Fig. 4.3). However, this is more of an

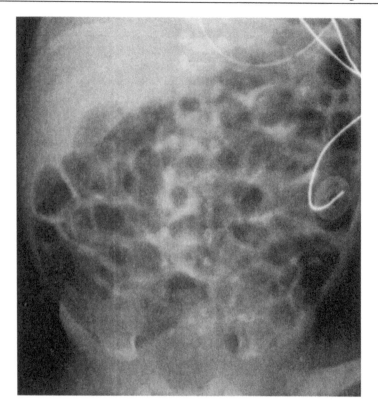

Fig. 4.1. Normal bowel gas pattern. The configuration, caliber and wall thickness are all typical.

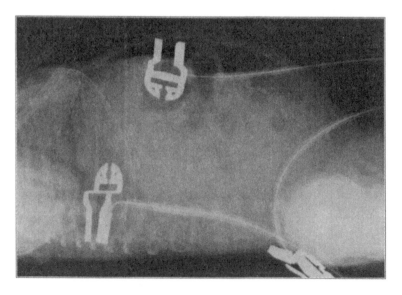

Fig. 4.2. Cross table lateral view illustrating clarity with which free air is identified beneath the dome of the anterior abdominal wall. A thick walled loop of bowel protrudes into the air collection.

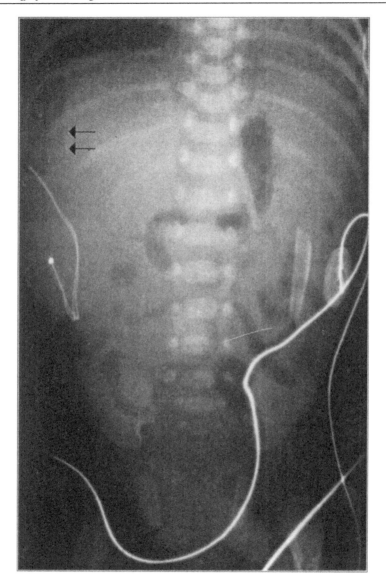

Fig. 4.3. Left lateral decubitus view showing air between the liver and the lateral abdominal wall (arrows). There is also increased separation of adjacent loops of bowel and narrowing of the lumen due to their thick walls. The loops of bowel are also straightened.

incursion on an ill baby and is more technically exacting. These infants are usually ventilated, therefore an erect view is not considered.

Once a NEC watch has been instituted supine and horizontal beam films are obtained every six to eight hours depending on the severity of the clinical signs and the findings on the preceding films. In our institution this status is maintained for 24-48 hours.

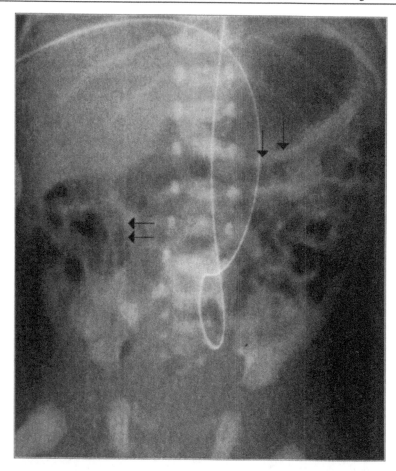

Fig. 4.4. Straightening or loss of the normal bowel configuration is seen in the right side of the abdomen and in the transverse colon (short thin arrows). A normal gas pattern is seen in the left lower quadrant. This child had clinical features of NEC which progressed.

Radiographic Features

The radiographic features include straightening of the bowel wall, dilatation of bowel, bowel wall thickening, intramural air, a fixed loop, free air, gasless abdomen, abdominal distension, and portal venous air.

Normally the abdomen has a pattern of slightly segmented lengths of air-filled bowel with concave inner margins (Fig. 4.4). The bowel wall curvature is approximately symmetrical on opposite sides. The wall is thin and best appreciated when two gas-filled loops are adjacent such that the combined thickness is 1 mm or less and an upper limit of approximately 1.5 mm. The caliber of the bowel has to be considered relative to other normal structures in the particular infant. A number cannot be assigned, as the children vary in size and gestational age, usually the normal bowel caliber is equal to a vertebra and a disc space or less. Upper limit of normal being a vertebra and two disc spaces. Occasionally normal colon may be a little larger than this. Typically the bowel contains a modest

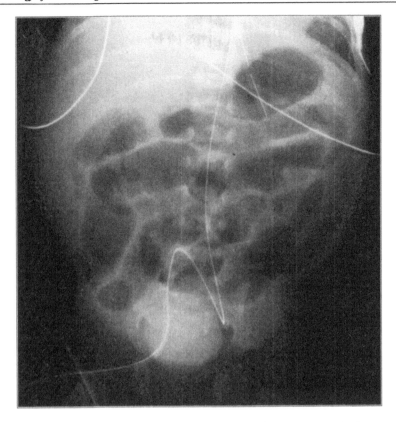

Fig. 4.5. Dilated bowel in a child with NEC (Contrast in the bladder from a prior procedure).

amount of gas. No gas or an excess of gas can occur with normal bowel but these extremes should prompt an evaluation for pathology.

Straightening of the bowel refers to a pattern where the normal segmented appearance disappears and a featureless length of bowel is seen (Fig. 4.4). Our neonatal intensivists describe a physical finding of "loopiness" which often corresponds to this radiographic feature.

Dilatation describes the increase in caliber so commonly associated with NEC. It is often accompanied by straightening (Figs. 4.5, 4.6). The significance of the increased caliber can be refined by gauging the amount of ballooning or tension in the demonstrated bowel. For example does it seem about to burst like an overblown balloon or is there a lot of slack in the wall?

Bowel wall thickening results from edema of the inflamed bowel. This is seen as an increase in the separation of adjacent air filled loops of bowel and may also be identified as indentation of the gas pattern equivalent to "thumb-printing" in adults (Figs. 4.7-4.9). There is loss of the regular concavity of the margins of gas-filled bowel and loss of the mirror image curvature of the opposing walls.

The cardinal sign of NEC is intramural air or pneumatosis intestinalis. This gas lies within the bowel wall (Fig. 4.10). It may appear as bubbles which may be as small as one or two millimeters or linear lucencies paralleling the bowel wall and luminal gas (Figs. 4.11, 4.12). It may be seen in profile or enface. The latter can be difficult to distinguish

Fig. 4.6. Preterm infant with a complicated cardiac history who developed NEC demonstrating dilated bowel loops. There is pneumatosis on the left.

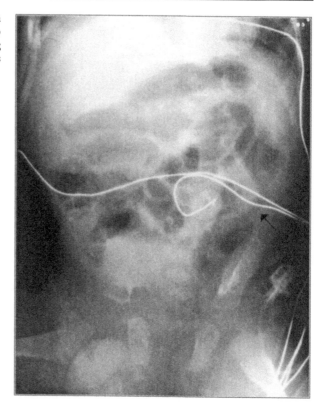

Fig. 4.7. Extensive bowel wall thickening (thick arrow) and thumbprinting or indention (thin arrow) of the bowel wall by edema. There is also free air outlining the left side of the falciform ligament (thin arrow).

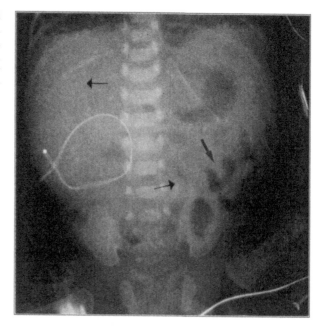

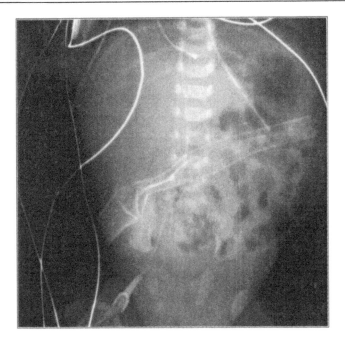

Fig. 4.8. Bowel wall thickening well seen between adjacent bowel loops. A Penrose drain was placed following perforation of the bowel.

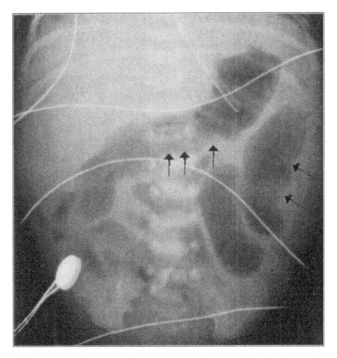

Fig. 4.9. Bowel wall thickening with increased distance between loops and thumbprinting in transverse and descending colon(arrows).

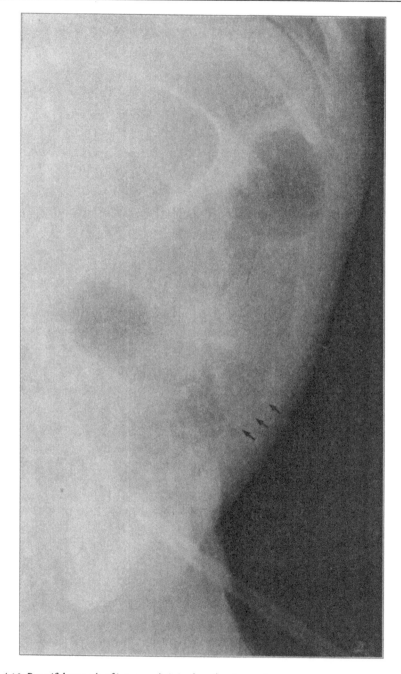

Fig. 4.10. Beautiful example of intramural air in the colon (arrows).

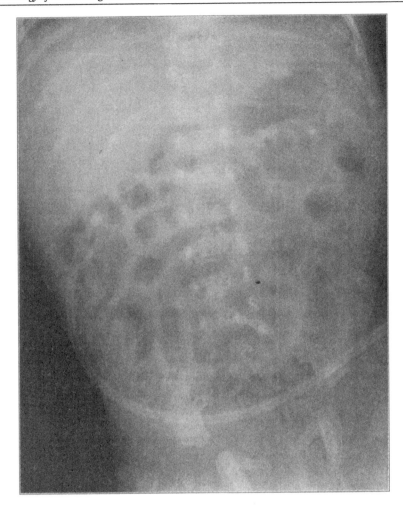

Fig. 4.11. Linear pneumatosis well seen on the left side of the abdomen. Bowel wall thickening is present virtually everywhere else.

from stool mixed with air. A follow-up study or a prone evaluation may help distinguish these by movement of stool in the one and redistribution of air in the other (Fig. 4.13). The quantity of gas ranges from little to extensive, but its presence has great diagnostic weight and therefore a careful search pays dividends.

A fixed loop describes the presence of an air collection which does not change in configuration on consecutive films. When present for more than 24 hours it is highly suspicious for a portion of bowel which has infarcted. This is a serious state and the bowel is at risk of perforation. Therefore this finding may precipitate surgical evaluation and possibly surgery.

Pneumoperitoneum indicates perforation of bowel which is a grave complication. The easiest feature to appreciate is air casting a profile of the external surface of bowel beneath the anterior abdominal wall on a cross-table lateral film (Fig. 4.2). The edge of the

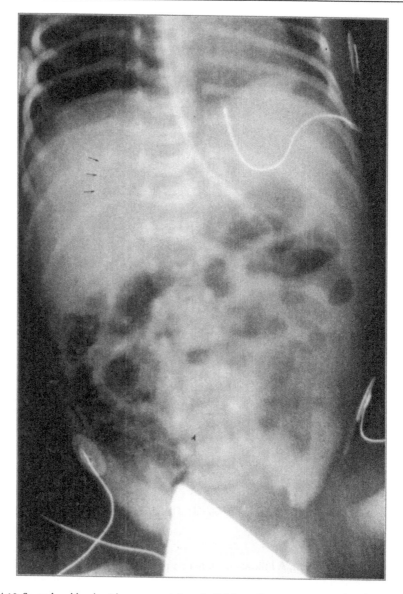

Fig. 4.12. Seven day old male with pneumatosis intestinalis bilaterally, more pronounced on the right. This is a bubbly pattern. There is free air outlining the left side of the falciform ligament (short thin arrows) and creating reduction in density on the left side of the liver. Retroperitoneal air is also present. See Figure 4.2 for companion cross-table lateral.

liver seen separated from the right lateral abdominal wall by air on a left lateral decubitus view is also easily detected on a satisfactory film (Fig. 4.3). Many other signs of free air have been described.

In the current context identification of the falciform ligament, as a linear soft tissue density in the right upper quadrant, with reduced density around it due to adjacent air on a supine abdominal film is one of the more commonly seen signs of free air. The air may

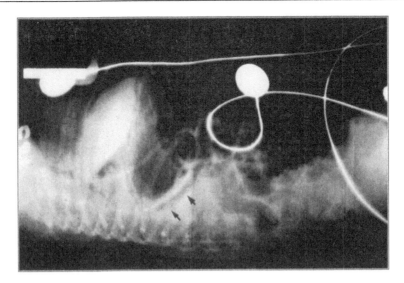

Fig. 4.13. Gastric pneumatosis due to NEC (arrows). There is also free air anterior to the liver.

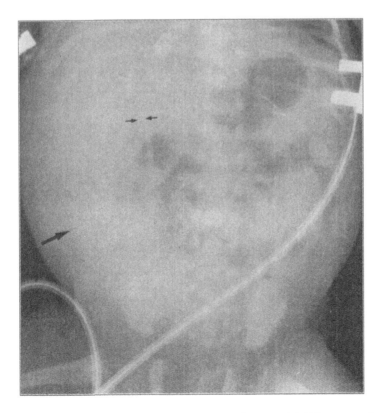

Fig. 4.14. Falciform ligament outlined by pneumoperitoneum. (short arrows) Bubbly pneumatosis seen in the right side of the abdomen (long arrows). Thumbprinting on the left.

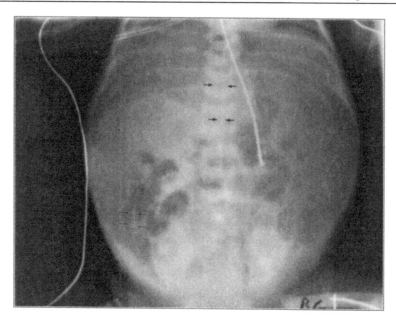

Fig. 4.15. Free air outling the falciform ligament (short fat arrows). Air collecting as a large lucency overlying most of the abdomen (football sign). Air on the inside and outside of the bowel wall (short thin arrows).

Fig. 4.16A. Twenty-three week gestation infant with suspected NEC showing a gasless abdomen.

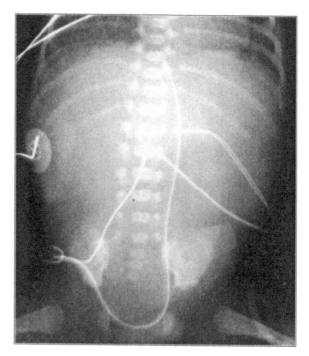

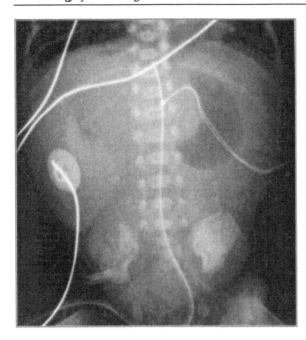

Fig. 4.16B. A small amount of air was added through the nasogastric tube filling the stomach prior to filming.

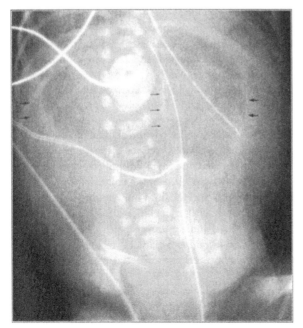

Fig. 4.16C. Twelve hours later there is clear demonstration of free air oulining the falciform ligament (short thin arrows). Also central air collection (football sign) (opposing short fat arrows).

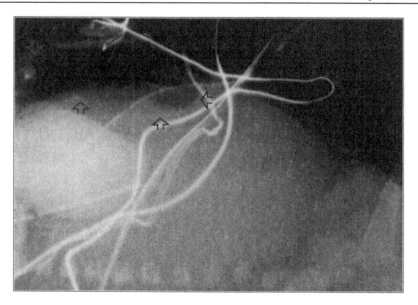

Fig. 4.16D. The companion cross-table lateral illustrates the free air anteriorly (open arrows) and the nasogastric tube lying in the gasless stomach.

collect as a large radiolucent central collection (football sign). Occasionally it is possible to see both the inside and outside of the bowel wall due to gas both inside and outside the wall (Figs. 4.14, 4.15).

A gasless abdomen on x-ray indicates no air in the bowel (Fig. 4.16A). In NEC the bowel may appear this way when filled with fluid. Good nasogastric suction may also result in a gasless appearance while a pharmaceutically paralyzed infant who cannot swallow may have a similar appearance. It can be deceptive. Neither bowel caliber nor thickness can be determined and perforation in this setting will not be appreciated as free air, but rather as free fluid not demonstrated on a radiograph. Some physicians will administer 2-3 ccs of air through a nasogastric tube in an attempt to reveal a perforation (Figs. 4.16B, 4.16C, 4.16D). Ultrasound evaluation of the abdomen may reveal free fluid.

Abdominal distension is recognized by bulging flanks, elevated diaphragms and tensely domed anterior abdominal wall. It may due to dilatated gas or fluid-filled bowel, ascites or body wall edema. Ascites may be suggested by gas-filled bowel collecting centrally and peripheral grayness due to bowel floating in a pool of fluid on a supine film. Clinical correlation will establish the presence of body wall edema. An US evaluation may be necessary to confirm fluid-filled dilatated bowel and ascites.

Portal venous air is recognized by linear lucencies radiating over the liver and often most pronounced peripherally in the terminal branches of the portal system (Fig. 4.17). It is an indicator of poor prognosis.

The constellation of radiographic features can give strong evidence of NEC. However these features must be weighed along with the clinical features to ensure accurate differential diagnosis. Other processes must be considered, for example: bowel dilatation due to obstruction or, more simply, copious swallowed air in a child on CPAP (Fig. 4.18), bowel wall thickening due to edema from heart failure, hemorrhage, hypoalbuminemia, or hydrops fetalis, intramural air due to enterocolitis complicating Hirschsprung's disease, or even the

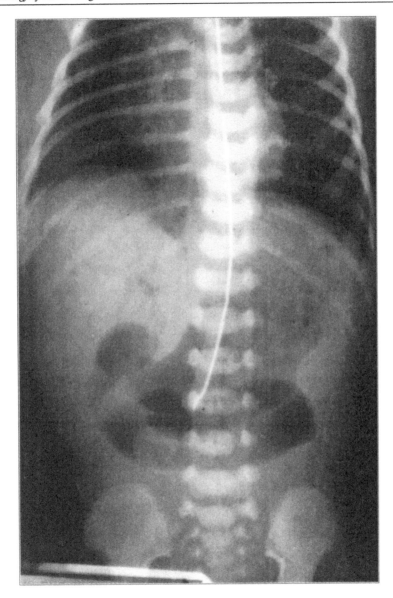

Fig. 4.17. Portal venous gas.

misinterpretation of air mixed with stool. (This is a not uncommon difficulty) Free air may arise from pneumomediastinum tracking into the abdomen or recent abdominal surgery.

Ultrasound

Ultrasound may illustrate some diagnostic signs serendipitously or be used to provide supportive evidence of abdominal disease. Bowel wall edema can be identified as thickening of bowel wall (Fig. 4.19). Increased bowel caliber may be seen well especially when fluid filled. The presence or absence of peristalsis and amount of bowel activity may be

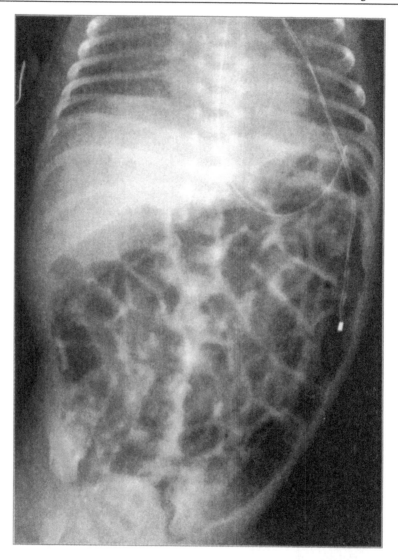

Fig. 4.18. CPAP belly. Extensively gas-filled bowel with normal configuration and wall thickness with caliber at the upper limit of normal.

assessed. Ascites is readily identified. When a large quantity is present the bowel floats in it. Intramural gas may be distinguished however this can be difficult when there is also intraluminal gas (Fig. 4.20). Portal venous gas is sometimes seen bubbling into the liver where it collects in the highest peripheral regions of the liver (Figs. 4.21, 4.22). Typically this is anterior in the supine patient. Therefore, the rest of the liver evaluation may need to be performed from a lateral or inferior approach to avoid obstruction by air shadowing. The air is identified as echogenic, linear collections conforming to vascular channels.

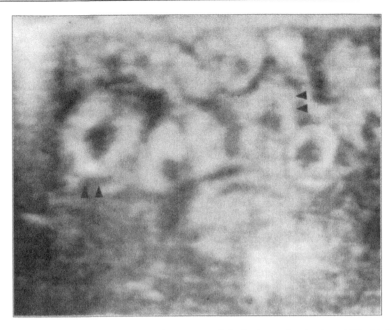

Fig. 4.19. Ultrasound evaluation of the bowel showing wall thickening and ascites in a child with NEC.

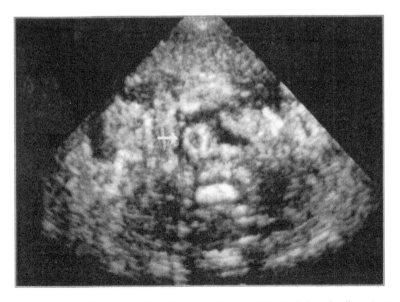

Fig. 4.20. Twenty-three week gestation infant with a ring of echogenicities in the bowel wall consistent with pneumatosis intestinalis. Sonolucent ascites was also present.

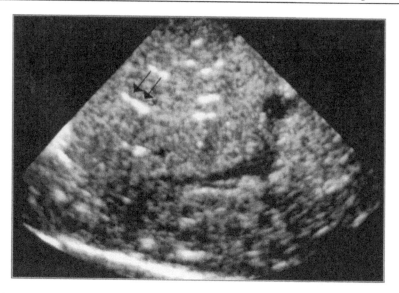

Fig. 4.21. Twenty-three week gestation infant on day two of life with echoic lines in the liver representing gas in the portal system. This bubbled through the vessels during the study (same child as in Fig. 4.20).

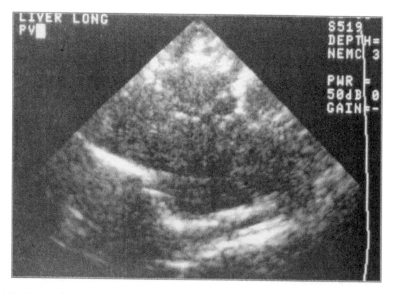

Fig. 4.22. Longitudinal view of the liver showing anteriorly sited echogenicities due to portal venous gas in a child with NEC.

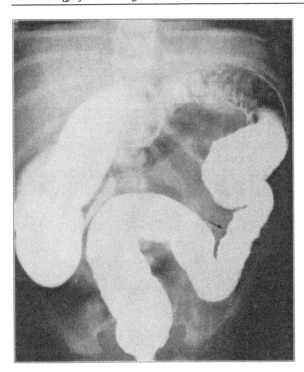

Fig. 4.23A. An enema was performed in the acute phase of illness demonstrating ulceration in the descending colon (arrow).

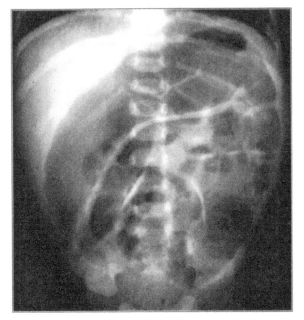

Fig. 4.23B. Three months later the infant had bowel obstruction with dilated bowel on the KUB.

Fig. 4.23C. An enema revealed a short stricture at the site that was previously ulcerated (between open arrows).

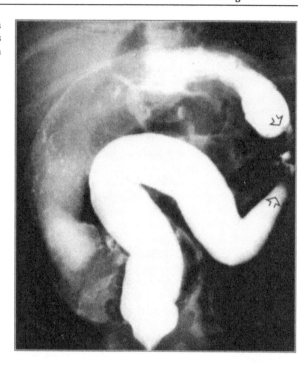

Fig. 4.24. Small bowel follow through on a 3 month old boy showing dilated bowel distally due to a NEC stricture (same child as in Fig. 4.12).

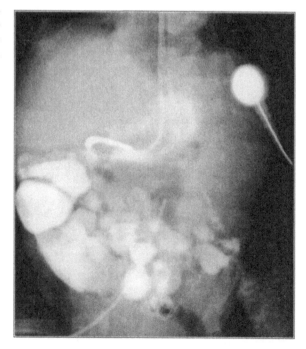

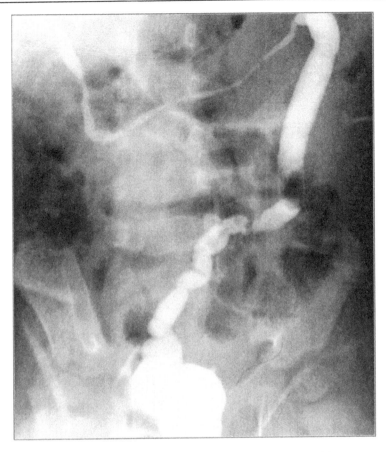

Fig. 4.25. Contrast evaluation of the mucus fistula demonstrates a long stricture of the transverse colon.

Complications

The radiological contribution to the acute process has been discussed. Now the role of the radiology service in management of complications and consequences of necrotizing enterocolitis will be reviewed.

Complications of necrotizing enterocolitis include abscess, stricture and fistula formation, development of an adynamic bowel loop and short gut syndrome.

Abscesses tend to present in the acute phase. US review of the abdominal and pelvic recesses together with assessment of the bowel for fixed masses and fluid collections is the method of choice for evaluation.

Strictures often become apparent when feeding is recommenced. Typically these are in distal small bowel or the colon. Investigation usually starts with an enema using a water soluble low osmolar contrast agent such as Isovue 128 (Figs. 4.23A, 4.23B, 4.23C). This is isotonic. If this fails to reveal an obstruction or stricture then a small bowel study is performed (Fig. 4.24). Strictures may be multiple, therefore if no caliber change is seen proximal to a stricture found on an enema then a small bowel study may also need to be performed.

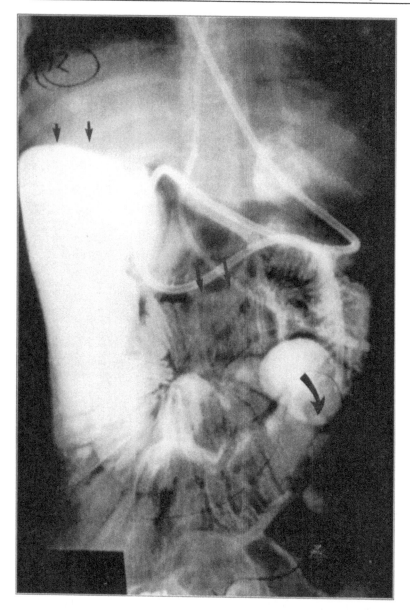

Fig. 4.26. Enteroclysis on a 10-year-old girl with short gut syndrome following NEC. Longer straight arrows demonstrate hypertrophied small bowel with increased caliber and normal valvulae conniventes and no wall thickening. This portion of the bowel had normal peristalsis. The very large caliber featureless loop on the right showed no peristalsis (short straight arrows). Contrast was encouraged through this segment by positioning the patient and using gravity to move the liquid. Beyond this segment contrast entered more hypertrophied small bowel which retained peristalsis before passing through the ileocolic anastomosis (curved arrow) to enter the residual colon (course of contrast distally marked by long-line on image). Following this procedure a fat breath analyzer test established a diagnosis of bacterial overgrowth presumed to be in this stagnant loop.

Children who have required an enterostomy and mucus fistula in the course of management generally have a contrast evaluation of the mucus fistula preferably in an antegrade fashion prior to reanastomosis to assess for strictures (Fig. 4.25). This is performed through the mucus fistula using a small calibre, soft catheter, such as a 5 FR red rubber. If a Foley catheter is used and it is decided to inflate the balloon, then this should be done cautiously under fluoroscopic guidance with as small a volume as possible to reduce the possibility of perforation of the narrow preterm bowel. If feeding has progressed satisfactorily the proximal bowel does not need evaluation before surgery.

Rarely fistulae form between adjacent diseased bowel loops. This may present with symptoms such as obstruction or infection. An antegrade contrast study with frequent imaging to assess the route contrast passes through the bowel will aid diagnosis of the site and nature of the problem.

Occasionally a late complication is malabsorption. This can be due to bacterial overgrowth in an adynamic bowel loop. This may arise when a compromised bowel loop does not recover fully and peristalsis is lost. A small bowel series or small bowel enema performed with close attention to the activity and caliber of all bowel loops may demonstrate such an inactive and possibly dilatated segment. A fat breath analyzer test may be performed to confirm bacterial overgrowth.

Short gut syndrome is a consequence of severe disease where a great deal of bowel has been resected. That which remains is hypertrophied so that its caliber is greatly increased. These patients may be perfectly well; however on plain film the bowel is dilatated and on contrast studies there are very few loops of dilated but otherwise normal bowel (Fig. 4.26).

Conclusion

The principal contribution of radiology to the diagnosis of NEC is in the assessment of standard radiographs for the various features that will suggest the diagnosis. Contrast studies are of greatest value in the evaluation of complications. US has a role in the diagnosis of acute disease as well as management of complications.

The radiologist needs to maintain great awareness to the possibility of the illness to enable prompt intervention.

CHAPTER 5

The Medical and ICU Management of Necrotizing Enterocolitis

Marc Lessin

A dvances in neonatal intensive care have lowered the limits of viability to as low as 22 weeks gestational age.[1] Great progress has been made in respiratory support with the use of surfactant, high frequency ventilation, and steroid administration. Necrotizing enterocolitis (NEC) is a disease associated with prematurity, and its prevalence has been increasing with advances in neonatal respiratory care. Survival of the extremely premature and small infant remains tenuous, however, with NEC claiming a significant mortality. Indeed, with continued advances in neonatal respiratory care, NEC may soon surpass chronic lung disease as the main morbidity and mortality of the premature infant.

Several theories have been advocated regarding the etiology of necrotizing enterocolitis. Cause and affect, however, have not been established. It is likely that its aggravating circumstances are multifactorial. The risk factors associated with NEC are well defined, but the mechanisms of injury are not fully elucidated. While principles of medical management are well-established, controversy exists as to the best treatment for the extremely premature or low birth weight infant with NEC, especially regarding the timing and indications for operative intervention and whether peritoneal drainage should be performed.

Epidemiology

The overall incidence of NEC is 0.1-0.3% based on live births and is more than one hundred times more prevalent in very low birth weight infants (1000 g).[2,3] In a widely quoted review the mean gestational age in 123 patients with NEC was 31 weeks. NEC occurs in most series in about 10% of premature infants admitted to the neonatal intensive care unit.[4]

Pathophysiology

The causative agent or etiology of NEC is not known. Several risk factors have been studied. Prematurity is associated with over 90% of known cases of NEC. The most consistent risk factors appear to be related to an immature gastrointestinal immunologic defense system. It is noteworthy that 11% of the patients with NEC have no identifiable risk factors.[4]

Newborn gut mucosa is deficient in secretory immunoglobulin A and B as well as T lymphocytes, especially in the non breast fed infant.[5] The deficiency of immunoglobulin A, particularly in the ileocolic area (the most common site of NEC) results in relatively unimpeded translocation of bacteria. Breast milk has been shown to have a protective

Necrotizing Enterocolitis, edited by Brian F. Gilchrist. ©2000 Eurekah.com.

effect, perhaps related to its inherent immunoglobulin.[6] In one study, the oral administration of immunoglobulins A and G was effective in preventing NEC in low birth weight infants, but in another study it was not preventative when compared to oral gentamicin which was also effective.[7,8]

Glucocorticoids given prenatally (to promote lung maturity) and postnatally have been shown to protect against necrotizing enterocolitis in low birth weight infants.[9,10] The mechanism suggested is promotion of intestinal mucosa maturation. This augments the mucosa's barrier function against bacterial translocation. Steroids' anti-inflammatory effect may also inhibit the cascade of mediators associated with progressive injury to the bowel wall.

The ischemia-reperfusion model was first introduced by Lloyd[11] citing the "diving reflex" as a cause for a low flow state. This theory was based on the shunting of blood away from the splanchnic circulation (as well as the skin and kidneys) to maintain blood flow to the heart and brain. This was thought to be analogous to the clinical situation of perinatal stress. The problem with this theory is that autoregulatory mechanisms exist in local vascular beds which respond by vasodilatation to acidosis and hypoxia. Secondly, most NEC occurs after one week, often without any documented physiologic stress to the newborn.

It is also notable that several studies have found an association with vasoactive medications. Indomethacin used to prevent premature labor has been implicated in NEC.[12] In addition, its use in closing a ductus arteriosus has pronounced vasoconstrictive effects on arterial beds, particularly in the splanchnic circulation.[13] Early surgical ligation of a patent ductus arteriosus has been associated with a decreased incidence of NEC.[14,15] Dopamine in high doses causes alpha-adrenergic stimulation which decreases splanchnic vascular flow. Both are used with frequency in the newborn nursery to support the ill, premature newborn. A true cause and affect between these medications and NEC is difficult to prove since the intestinal insult may be a coincident event due to the overall severity of illness.

Cocaine is considered an etiologic factor in the development of NEC. Cocaine is a potent vasoconstrictor which may decrease placental blood flow resulting in fetal hypoxia. In one study there was a 2.5-fold increase in developing NEC in infants born to mothers who used cocaine during their pregnancy. Cocaine-positive infants with NEC required surgery more often and had a higher mortality in comparison to matched controls with NEC.[16]

Multiple infectious organisms have been identified but no single specific organism has been implicated. Two patterns of NEC have been noted: a sporadic type and an epidemic outbreak. In the sporadic type, often normal gut flora is found; however, pathogenic strains of *Escherichia coli*, *Klebsiella pneumoniae* and *Staphylococcus epidermidis* are commonly identified. The epidemic type of NEC which is usually endemic in a given neonatal intensive care unit is usually due to a single inciting organism.[17] *Clostridium perfringens* has been associated with especially severe cases of NEC. Fungus has been implicated as the source of peritonitis in as many as 35% of NEC cases.[18] It is postulated that the delayed feeding schedule of premature infants, the need and low threshold for broad spectrum antibiotic usage, and the quasisterile environment of the neonatal intensive care unit delays the normal colonization of the protective gut flora.

Feeding has long been associated with the development of NEC. Ninety percent of NEC babies have been fed; 7-10% have never feed fed.[19] Most develop problems after initiation or advancement of feeds, usually shortly after the first week of life. Hyperosmolar feeds have been implicated in one study in which 87% of preterm infants fed on hyperosmolar formula developed NEC as opposed to 25% fed a low osmolality milk-based formula.[20]

Cytokines have been implicated in the pathogenesis of NEC. Endotoxemia mediates the release of platelet activating factor which causes a pattern of intestinal injury similar to NEC in animal models.[21] Speculation as to the role of nitric oxide as a mediator of injury has produced conflicting reports showing that it was preventative in one study and exacerbative in another.[22,23] Other cytokines implicated are tumor necrosis factor, interleukin-6 and oxygen free radicals. The role of cytokines as agents of tissue injury or for prevention of the associated microvascular injury associated with NEC are 0incompletely understood and have no clinical application at this time.

Diagnosis

The diagnosis of NEC is often made on a clinical basis supported by typical radiographic findings. The typical clinical findings are abdominal distension, feeding intolerance, abdominal wall cellulitis or meconium staining, abdominal tenderness and blood in the stool. The newborn may appear lethargic, have episodic bradycardia or apneic spells and temperature instability. Laboratory values may show metabolic acidosis, an elevated or depressed white blood cell count and thrombocytopenia. Hypoglycemia may be present from failure to absorb and metabolize enteric feeds. If the infant is receiving parenteral nutrition, unexplained hyperglycemia may herald sepsis. Radiographic findings are covered elsewhere, but a point worth mention in this monograph is that no findings are specific for NEC, except for pneumatosis intestinalis which is considered pathognomonic.

Classification

Several staging systems have been proposed to delineate the severity of clinical status and determine appropriate therapy. The most frequently used system is that proposed by Bell.[24] His system divides NEC into suspected (without radiographic findings), definite (with radiographic findings), and advanced (systemic manifestations) (Table 5.1). As one would expect, the more advanced the stage, the higher the incidence of bacteremia and mortality. The classification has mainly been useful for the inter-institutional study of NEC.

Treatment

Once the diagnosis for NEC is confirmed or suspected, feedings are immediately stopped. A sump type decompression gastric tube is placed to low continuous suction. Blood cultures are obtained prior to initiation of antibiotic coverage. Antibiotic coverage is directed to cover enteric organisms including anaerobes which are known to be present in the colonic gut flora after 48 hours of life. A strong index of suspicion for fungal infection must be maintained in the patient who fails to show improvement or in whom bacterial cultures remain negative in an otherwise symptomatic patient. Total parental nutrition is instituted once the child is fluid resuscitated and stabilized. This requires central venous access to meet caloric requirements. Serial abdominal x-rays are obtained every 8 hours initially to look for signs of progression of disease and specifically to detect complications associated with NEC, such as free air. Cross-table lateral or left lateral decubitus films are necessary since the detection of free air may be difficult on the flat plate film. Frequent physical examination of the abdomen is performed, preferably by the same examiner (Table 5.2).

Table 5.1. NEC staging system[24]

Stage 1 (Suspected NEC):
1. Temperature instability, lethargy, apnea or bradycardia.
2. Poor feeding, increased gastric residuals, emesis (bilious or positive for occult blood), mild abdominal distension, occult blood in the stool.
3. Radiographs show mild ileus.

Stage II (Definite NEC):
1. Symptoms as in Stage 1 plus persistent occult or gross gastrointestinal bleeding, marked abdominal distension.
2. Radiographs as in Stage I plus bowel wall edema or ascites, fixed intestinal loops, pneumatosis intestinalis or portal venous gas.
3. Laboratory values may show metabolic acidosis, thrombocytopenia, leukocytosis or leukopenia.

Stage III (Advanced NEC):
1. Symptoms as in Stage I plus Stage II plus hemodynamic changes, evidence of septic shock or significant gastrointestinal hemorrhage.
2. Additional radiographic finding may demonstrate pneumoperitoneum.
3. Laboratory values as in Stage II plus disseminated intravascular coagulopathy (DIC).

Table 5.2. Medical management of NEC

1. Cessation of enteral feeds.
2. Cultures obtained from blood, urine, sputum and CSF if indicated.
3. Antibiotics to begin after cultures obtained. If already on antibiotics, adjust to cover enteric organisms including enterococcus, gram negative enteric pathogens and anaerobic colonic flora.
4. Gastric decompression on low continuous suction with a sump type tube.
5. Serial abdominal examinations preferably by the same examiner.
6. Radiographs of the abdomen including cross table lateral every 6-8 hours for the first 48 hours.
7. Initial fluid resuscitation to correct hypovolemia.
8. Total parenteral nutrition preferably with a centrally located intravenous catheter once fluid resuscitation is complete.

NEC is a disease which is successfully treated over 80% of the time medically. Kosloske evaluated 12 criteria as indicated for laparotomy based on their predictive value of intestinal gangrene. She found the best indicators were pneumoperitoneum, portal venous gas and organisms found on paracentesis.[25] I have found pneumoperitoneum the only reliable indicator, and even this can be secondary to non-gastrointestinal sources such as dissection from an associated pneumothorax or gastric perforation from a nasogastric tube. Portal venous gas is associated with extensive intestinal necrosis, but does not in itself indicate that celiotomy is necessary. Early operation for portal venous gas in an unstable patient may result in improved survival.[26] Paracentesis, while indicative of perforation, can have false positive findings due to intestinal puncture. The Editor uses only free air and failure to improve with maximal medical therapy as indications for surgery.

Peritoneal Drainage

The treatment of pneumoperitoneum associated with NEC was first reported from Toronto in 1997.[27] Since this initial report, peritoneal drainage has evolved from its original application to neonates too ill to undergo laparotomy to a definitive therapeutic intervention especially in the extremely low birth weight infant. The Toronto group found that one third of patients drained required no intraabdominal operative intervention whatsoever. The remaining two thirds fell roughly into three equal categories. One third had a rapid irreversible fatal course which could not have been averted even with surgery. The second group failed to show improvement and required operative intervention with a 50% mortality. The final group developed late complications after recovery from their NEC such as bowel obstruction due to strictures, fistulas or hernias at the drain site. This group with late complications was well recovered from the initial septic episode and had gained significant weight making the operative intervention much less risky. Survival with peritoneal drainage alone is comparable or better than that with laparotomy in roughly matched control groups with pneumoperitoneum.[27-32]

I advocate placement of a peritoneal drain (Penrose) with irrigation of the abdominal cavity through a small incision in the right lower abdomen for pneumoperitoneum, especially in neonates weighing less than 1000 g. Cultures are sent from the peritoneal effluent. This procedure can be done at the bedside under local anesthesia with sedation without transport to the operating room. Antibiotic coverage is continued, along with gastric decompression and total parenteral nutrition. Drains are removed after about one week when there has been cessation of drainage, and clinical improvement. Feeds are initiated slowly after 10-14 days.

The successful treatment of intestinal perforation with peritoneal drainage is not well understood. Well-established surgical principles dictate that perforations of the intestine require exploration with removal of all necrotic intestine and exteriorization of proximal normal intestine. In the extremely low birth weight infants, the combined effects of general anesthesia and major abdominal surgery increase the risk of hemodynamic instability caused by hypotension, transfusion requirements, third spacing of fluids and hypothermia. The small premature newborn has a poorly understood mechanism of scarless gastrointestinal healing and immunity that may prevent overwhelming sepsis that one would expect from gross peritoneal contamination.

Peritoneal drainage may be considered definitive therapy as long as there is rapid improvement in the patient's condition. If not, it is recommended that the patient be taken for celiotomy. A vexing question remains regarding how long peritoneal drainage should be maintained before proceeding to the operating room. In my experience, it has occasionally taken several days to show improvement in some who have ultimately survived. The initial indication for peritoneal drainage was in premature newborns who were too ill to proceed to surgery. The paradox is that if drainage rather than laparotomy was chosen initially because of the infant's poor condition and low weight, worsening clinical status after drainage hardly makes the patient a better operative candidate, and "last effort" celiotomy might reasonably be omitted.

Recovery Phase

Once the patient has recovered from NEC, feeds should be resumed cautiously. Generally at least 7-10 days will have elapsed since the initial diagnosis or drain placement. Feeds are begun via a gavage tube in the stomach since most will be too small to take feeds by mouth. Small amounts of half-strength elemental formula or preferably breast milk are administered. Careful monitoring of the baby for any signs of feeding intolerance are noted. Abdominal examinations for distension are performed frequently. Evidence of gastric residuals in excess of the infusion, abdominal distension or vomiting requires discontinuation of feeds. Often the feeds need to be spaced over long intervals as the recovering gut adjusts to the absorptive load. Feedings are progressed to full volume and concentration over the next 7-10 days.

Late Complications

Successful medical management of NEC as well as peritoneal drainage (or laparotomy) have been associated with stricture formation, particularly in the distal colon. In the recovery phase of NEC, development of feeding intolerance, unexplained sepsis or persistent blood in the stool should raise the suspicion of stricture. This should be evaluated with contrast studies.

Recurrent NEC has been an infrequent occurrence. In a study from England, it occurred in 6%, and at a median of 37 days after the initial episode. The great majority (70%) were treated medically.[33]

Neurologic compromise remains a significant problem for those recovering from NEC, particularly after operative intervention. A study of 40 survivors of NEC compared to premature infants without NEC showed no worse neurologic outcome; however, only 48% were developmentally normal.[34] Low birth weight premature infants with advanced disease (Bell Stage III), 75% of whom required surgery for perforated NEC, had delayed growth with decreased head circumferences and 43% had severe neurodevelopmental impairment at two year follow-up. This, however, may have been related to the overall severity of their disease rather than the NEC.[35]

Future Directions

The most effective preventative measure against NEC is the prevention of premature delivery. In countries where premature birth rates are low such as Japan, there is a notable decreased incidence of NEC.[36] Readily available prenatal care with aggressive management of premature labor is probably the most important and controllable factor in the prevention of NEC.

The administration of glucocorticoids should be routine when premature labor is likely to result in early delivery. The benefit to maturation of the respiratory system is well established and several studies have shown their protective effect against NEC of the gastrointestinal tract. Their administration in the postnatal period has been shown to be helpful but has potential adverse side effects especially related to the already assaulted immune system of the premature neonate.

The use of breast milk for initiation of feeds should be strongly encouraged whenever possible, and the administration of pasteurized donated breast milk may provide valuable sources of immunoglobulin rich infusion. When breast milk is not administered, feeds should be started cautiously, advanced slowly, and be hyposmolar.

Indomethacin should be used cautiously and in low doses. If the child has any signs of NEC it should be avoided or discontinued. If the child has ductal related hemodynamic and respiratory compromise, early surgical ligation may be the preferred option. In neonates requiring vasopressor support, dopamine should be kept at low doses due to its vasoconstrictive action on the splanchnic circulation at higher doses. Dobutamine is equally effective in maintaining hemodynamics, so long as the patient is euvolemic. Dobutamine increases myocardial contractility while decreasing afterload, which should have less detrimental effects on mesenteric blood flow.

The role of cytokines as inflammatory mediators is incompletely understood at this time. Further research requires their identification and subsequent interventions to mediate their expression where appropriate. The role of cytokines, however, has no clinical application presently.

The role of peritoneal drainage is increasing and should probably be the initial therapy for pneumoperitoneum especially in infants less than 1000 g. Multiple retrospective reports show that survival with peritoneal drainage alone is as good as (and probably better) than laparotomy. Its critics point out that it fails to remove a potential source of continuing sepsis.[37] I, however, believe that the extent of disease is increased by the laparotomy itself due to local vasoconstrictive effects brought on by hypothermia, hypoxia, hypovolemia and acidosis. In addition, peritoneal drainage has not been associated with the severe sequelae of short bowel syndrome which may result after extensive intestinal resection. The successes and benefits of peritoneal drainage cannot be discounted, and its applications may well become the mainstay in specific situations particularly in the extremely low-birth-weight premature infant.

References

1. Hack M, Friedman H, Fanaroff AA. Outcomes of extremely low birth weight infants. Pediatrics 1996; 98:931-937.
2. Holman RC, Stehr-Green JK, Zelasky MT. Necrotizing enterocolitis mortality in the United States. Am J Public Health 1989; 79:987-989.
3. Stoll BJ. Epidemiology of necrotizing enterocolitis. Clinics Perinatology 1994; 21:205-218.
4. Kliegman RM, Fanaroff AA. Neonatal necrotizing enterocolitis: A 9 year experience. Am J Dis Child 1981; 135:603-608.
5. Udall JN. Gastrointestinal host defense and necrotizing enterocolitis: An update. J Pediatr 1990; 117:S33-S44.
6. Lucas A, Cole TJ. Breast mild and neonatal necrotizing enterocolitis. Lancet 1990; 336:1519-1523.
7. Eibl MM, Wolf HM, Furnkranz H et al. Prevention of necrotizing enterocolitis in low-birth-weight infants by IgA-IgG feeding. N Engl J Med 1988; 319:1-7.
8. Fast C, Rosegger H. Necrotizing enterocolitis prophylaxis: Oral antibiotics and lyophilized enterobacteria vs oral immunoglobulins. Acta Pedatr 1994; 83(Suppl 396):86-90.
9. Halac E, Halac J, Begue EF et al. Prenatal and postnatal corticosteroid therapy to prevent neonatal necrotizing enterocolitis: A controlled trial. J Pediatr 1990; 117:132.
10. Bauer CR, Morrison JC, Poole WK et al. A decreased incidence of necrotizing enterocolitis after prenatal glucocorticoid therapy. Pediatrics 1984; 73:682-688.
11. Lloyd JR. The etiology of gastrointestinal perforation in the newborn. J Pediatr Surg 1969; 4:77-84.
12. Norton ME, Merrill J, Cooper BAB et al. Neonatal complications after the administration of indomethacin for preterm labor. N Engl J Med 1993; 329:1602-1607.
13. Cronen PW, Nagaraj HS, Janik JS et al. Effect of indomethacin on mesenteric circulation in mongrel dogs. J Pediatr Surg 1982; 17:474-478.
14. Cassady G, Crouse DT, Kirklin JW et al. A randomized, controlled trial of very early prophylactic ligation of the ductus arteriosus in babies who weighed 1000 g or less at birth. N Engl J Med 1989; 320:1511-1516.
15. Mikhail M, Lee W, Toews W et al. Surgical and medical experience with 734 premature infants with patent ductus arteriosus. J Thorac Cardiovasc Surg 1982; 83:349-357.
16. Czyrko C, Del Pin C, O'Neill J et al. Maternal cocaine abuse and necrotizing enterocolitis: Outcome and survival. J Pediatr Surg 1991; 26:414-421.
17. Moomjian AS, Packham GJ, Fox WW et al. Necrotizing enterocolitis: Endemic versus epidemic form. Pediatr Res 1978; 12:530.

18. Karlowicz MG. Risk factors associated with fungal peritonitis in very low birth weight neonates with severe necrotizing enterocolitis: A case-control study. Pediatr Infect Dis J 1993; 12:574-577.
19. Marchildon MB, Buck BE, Abdenour G. Necrotizing enterocolitis in low-birth-weight infants fed an elemental formula. J Pediatr 1975; 87:602-608.
20. Book LS, Herbst JJ, Atherton SO et al. Necrotizing enterocolitis in low-birth-weight infants fed an elemental formula. J Pediatr 1975; 87:602-608.
21. Sun XM, Hsueh W. Bowel necrosis induced by tumor necrosis factor in rats in mediated by platelet activating factor. J Clin Invest 1988; 81:1328-1331.
22. MacKendrick W, Caplan M, Hsueh W. Endogenous nitric oxide protects against platelet activating factor-induced injury in the rat. Pediatr Res 1993; 34:222-228.
23. Ford HR, Watkins S, Reblock K, Rowe M. The role of inflammatory cytokines and nitric oxide in the pathogenesis of necrotizing enterocolitis. J Pediatr Surg 1997; 32:275-282.
24. Bell MJ, Ternberg JL, Feigen RD. Neonatal necrotizing enterocolitis. Ann Surg 1978; 187:1-7.
25. Kosloske AM. Indications for operation in necrotizing enterocolitis revisited. J Pediatr Surg 1994; 29:663-666.
26. Grosfeld JC, Cheu H, Schlatter M. Changing trends in necrotizing enterocolitis. Ann Surg 1991; 214:300-307.
27. Ein SH, Marshall DG, Girvan D. Peritoneal drainage under local anesthesia for necrotizing enterocolitis. J Pediatr Surg 1977; 12:963-967.
28. Morgan LJ, Shochat SJ, Hartman GE. Peritoneal drainage as primary management of perforated NEC in the very low birth weight infant. J Pediatr Surg 1994; 29:310-315.
29. Lessin MS, Luks FI, Wesselhoeft CW et al. Peritoneal drainage as definitive treatment for intestinal perforation in infants with extremely low birth weight (< 750 g). J Pediatr Surg 1998; 33:370-372.
30. Cheu HW, Sukarochana K, Lloyd DA. Peritoneal drainage for necrotizing enterocolitis. J Pediatr Surg 1988; 23:557-567.
31. Ein SH, Shandling B, Wesson D et al. A 13-year experience with peritoneal drainage under local anesthesia for necrotizing enterocolitis perforation. J Pediatr Surg 1990; 25:1034-1037.
32. Azarow KS, Ein SH, Shandling B et al. Laparotomy or drain for perforated necrotizing enterocolitis: who gets what and why? Pediatr Surg Int 1997; 12:137-139.
33. Stringer MD, Brereton RJ, Drake DP et al. Recurrent necrotizing enterocolitis. J Pediatr Surg 1993; 28:979-981.
34. Stevenson DK, Kerner JA, Malachowski N et al. Late morbidity among survivors of necrotizing enterocolitis. Pediatrics 1980; 66:925-927.
35. Walsh MC, Kliegman RM, Hack M. Severity of necrotizing enterocolitis: influence of outcome at 2 years of age. Pediatrics 1989; 84:808-814.
36. Shimura K. Necrotizing enterocolitis? NICU 1990; 3:5.
37. Kosloske AM. Surgery of infants and children. In: Oldham KT, Colombani PM, Foglia RP, eds. Lippincott-Raven, 1977:1201-1213.

CHAPTER 6

The Surgery of Necrotizing Enterocolitis

Colin A.I. Bethel

Preparation of the Infant for Surgery

The overall medical condition of the infant with necrotizing enterocolitis who requires operation must be optimized prior to surgery. These infants frequently are in shock. Adequate intravenous lines should be established, and the infant should be given fluid to restore the intravascular volume to optimal levels. An appropriately-positioned umbilical artery catheter or a peripheral arterial line can be of great help in the management of the infant. The urine output should be 1.5-2.0 cc/kg/h. This can be achieved using crystalloid, packed cells, and fresh frozen plasma (FFP), as indicated by hematologic indices. Packed cells, FFP and platelets should be ordered and ready for use during surgery. The patient should be started on broad spectrum antibiotics to include coverage of gram negative enterics and anaerobic bacteria, if these have not already been started. The patient's oxygenation and ventilation should be optimized: with few exceptions all infants will be endotracheally intubated in the newborn intensive care unit; hypoxia and hypercarbia worsen the patient's overall tolerance of anesthesia, and perhaps the course of the necrotizing enterocolitis itself. These should be corrected prior to moving to the operating room. The patient should be maintained under a radiant warmer in order to avoid hypothermia. A dose of gammaglobulin may be a benefit, and should be considered.

General Operative Approach

These same considerations apply once the infant has been moved to the operating room. The infant should only be moved onto the operating table, if the room has been warmed appropriately. The infant should have a warming hat and be covered with blankets and/or plastic drapes until the last moment prior to starting the surgical prep. Care should be taken to make sure that an adequate airway has been maintained: it is paramount that the endotracheal tube be well-positioned above the carina prior to covering the infant with heavy drapes. In particular, the chest should be observed for adequate chest wall rise during inspiration. The infants should be monitored using transcutaneous oximetry, end-tidal capnography, and arterial and/or venous blood gases. The surgical prep should be performed expeditiously using warmed prep solutions in order to minimize hypothermia. The prep should be kept as dry as possible to avoid fluid pooling under the infant.

A transverse abdominal incision should be used. This incision is usually made in a right supraumbilical location, as it provides ready access to the ileocolic region which is involved in the majority of cases. The incision may be tailored given specific physical or

Necrotizing Enterocolitis, edited by Brian F. Gilchrist. ©2000 EUREKAH.COM.

radiologic findings, such as a left-sided mass on physical exam, or a fixed loop on serial radiographs. Wide exposure is the key to proper exploration and operation.

The infant is likely to have a coagulation abnormality from a combination of thrombocytopenia, disseminated intravascular coagulation (DIC), and hypothermia. Therefore extreme care must be taken to minimize bleeding. This may include the use of cautery dissection as much as possible, and the use of topical agents such as thrombin spray for control of diffuse oozing from raw, inflamed surfaces. Liver hemorrhage in the absence of iatrogenic injury has been described in low birthweight patients who required large volumes of preoperative fluid to correct hypotension. Spontaneous bleeding from the liver is associated with a 13% survival after surgery, in comparison to an 88% survival in the absence of liver bleeding. The surgeon should be aware of this potential pitfall; the use of inotropic agents in combination with earlier surgical intervention has been suggested to be of potential benefit in avoiding this problem.[1]

The operation should be carried out expeditiously in order to avoid hypothermia and anesthesia-related complications. The most expedient surgical approach is best, especially if the infant is unstable. The duration of total bowel evisceration should be kept to a minimum in order to keep hypothermia at bay. Hypothermia may be further minimized by keeping exposed bowel covered with warm moist sponges or drapes.

The first observation which may be made is the quality of free intraperitoneal fluid. Bilious material, blood, or brown fluid signifies the presence of intestinal necrosis or intestinal perforation. Cultures should be taken. The intestine is then inspected. Obviously involved loops of bowel are noted, but the entire bowel must be inspected. The stomach and duodenum are inspected for NEC involvement. If portions of the duodenum are involved, a complete Kocher maneuver of the duodenum may be indicated. Again, this maneuver should be performed using careful cautery dissection as much as possible to minimize time loss and bleeding from the retroperitoneum. The small bowel should be inspected and palpated from the ligament of Treitz to the ileocecal valve. The large bowel should also be examined closely. In cases in which the large bowel is involved, the lateral peritoneal attachments of the ascending and descending colons may need to be taken down, realizing that care should be taken not to injure the ureters which may be closely involved with an involved segment of bowel. Again, cautery dissection, topical thrombin spray, and frequent irrigation of the field with small volumes of warm saline to clear the field of debris, opaque fluid and blood are helpful.

The intestine is assayed for viability. If the bowel is white, full thickness necrosis has occurred. The distinction between viability and necrosis may be more difficult in intestinal loops which appear purple or dark green, because similar color changes may be expected in viable bowel with intramural hemorrhage. Palpation of these loops may be helpful because it may distinguish areas of bowel which appear on the surface to be potentially viable, but which are in reality paper thin and about to perforate.

Specific Considerations: Management of Focal Necrosis

In order to minimize the risk of short-gut syndrome, only perforated or necrotic segments of intestine should be resected. The ileocecal valve should be preserved, if possible.

NEC Involving a Single Segment

If a single necrotic segment is found, this segment is resected. There are several options available to handle the proximal and distal ends of remaining bowel. In most cases, the proximal end is brought out as a stoma. The distal end may be brought out as a mucous fistula, or simply closed and dropped back into the abdomen; in the latter situation, the distal limb should be marked with a clip or suture in order to facilitate its identification at the time of stoma closure. Reanastomosis can usually be accomplished in 4-6 weeks, but a contrast study of the colon and/or distal bowel through the mucous fistula should be performed in order to rule out distal strictures prior to closure.

The "clip and drop-back" technique has been described as an alternative to stoma formation, and provides a means of avoiding the fluid and electrolyte losses, intestinal fistulas and stenoses, and skin excoriation associated with high intestinal fistulae. Both proximal and distal bowel ends are simply clipped using a large hemoclip and dropped back into the peritoneal cavity; 48-72 hours later, after the infant has been allowed to stabilize, a primary anastomosis is performed.[2] The editor of this book strongly suggests a different approach.

Another alternative to stoma formation is performance of a primary anastomosis at the time of the initial exploration and bowel resection. In a prospective study of 18 infants treated in this fashion, there was an 11% mortality rate, a 22% incidence of recurrent NEC, and a 17% incidence of intestinal strictures. The authors concluded that these rates compared with those reported for staged operations.[3] Primary anastomoses should only be performed when the infant is hemodynamically stable, normothermic, and oxygenating and ventilating without difficulty as construction of an anastomosis with will prolong the operation.

NEC Involving Multiple Discrete Segments

The situation becomes much more complicated, if multiple segments of bowel are involved. In this setting, all the necrotic and/or perforated segments are resected, keeping in mind the importance of preserving bowel length. This can be accomplished by en bloc resection of multiple segments which are separated by very short bridges of intervening normal bowel, or by separate segmental resections. The most proximal segment of intestine should be brought out as a completely diverting end stoma, and the distal segments may also be brought out as several individual stomas. If, as is often the case, the bowel segments are too short to reach the abdominal wall, or the mesentery has been fore-shortened by the inflammatory process thereby prohibiting the bowel loops to be brought out as stomas, several anastomoses between the remaining distal loops of intestine may be performed. These will be protected by the diverting proximal stoma. If the patient's overall condition does not permit the anesthesia time required to perform multiple anastomoses, the use of a multiple-segment "clip-and-drop" technique in which the viable segments left in situ are clipped at each end and left for reanastomosis 48-72 hours later has been successfully used by this author.

The timing of stoma takedown depends on several factors. In general, restoration of intestinal continuity has been left for at least 6-8 weeks after the initial exploration. Some infants may develop stoma-related complications such as fluid losses, stomal stenosis, or abdominal wall excoriation; these may become so difficult to manage that early stoma closure may be in the best interest of the child. In a retrospective analysis of NEC patients, the complication rate for closure prior to 3 months after initial enterostomy was the same as for delayed stoma closure.[4] Weber has advocated early stoma closure, but points out that anastomotic dehiscence is more likely to occur if the infant has had poor weight gain or has a low serum albumin.[5]

Specific Considerations: Management of NEC "Totalis"

Up to 20% of infants have NEC involving more than 75% of the intestine. When the entire midgut appears nonviable, several treatment options are available.

Closure of the Abdomen and Withdrawal of Support

The abdomen may simply be closed with a plan to provide comfort measures for the patient prior to expiration. The family should be involved in the discussions leading to this decision. The editor convenes a meeting with the family, their clergyman and the hospital ethicist while the patient is in the operating room. This decision must be addressed in a timely and intimate manner.

High Jejunostomy in the Absence of Distal Perforation

A high stoma may be created using proximal jejunum. The distal bowel may be gently irrigated and suctioned out to help decompress the lumen, and should be brought out as a mucous fistula. High proximal enterostomy may facilitate recovery of some or all the distal intestine by: 1) decompressing the bowel and thereby decreasing intraluminal pressure and improving intestinal perfusion, 2) decreasing the intestinal metabolic rate and 3) decreasing the exposure to intraluminal inflammatory mediators, and bacterial byproducts. It also allows some time for the infant to be returned to the controlled environment of the intensive care unit where further optimization of hemodynamic and ventilatory status may take place.

The optimal location of the distal bowel mucous fistula on the abdominal wall has been the topic of controversy. Many surgeons prefer to locate the mucous fistula in a separate location from the jejunostomy. However, the mucous fistula may be brought out through the same incision as the jejunostomy, or even through the abdominal incision itself. The wound infection rate is not increased using this approach, and it may be helpful in decreasing the operative time in unstable infants. As a practical point, because the mesentery is often foreshortened, the stomas must be located wherever they can be made to reach the abdominal wall without tension.

If the infant becomes hemodynamically stable, is able to be weaned from pressor support, and has improved ventilatory parameters and signs of resolution of sepsis, the distal bowel may be presumed to have recovered to some degree. These infants should be placed on central hyperalimentation for 1-2 months with a plan to perform a contrast study through the mucous fistula and through the rectum to determine intestinal continuity. At 6-8 weeks, further exploration to resect strictures, can be performed, leaving the proximal jejunostomy in place as a means of continuing fecal diversion until the anastomoses have healed.

If the infant deteriorates or fails to improve over the next 48-72 while being monitored in the intensive care unit treatment should either be withdrawn, or the infant should be returned to the operating room for a "second look" procedure during which time necrotic intestine can be resected. A decision may be made at the time of the "second-look" procedure to withdraw support. This possibility highlights the importance of thorough and realistic parental counseling when pan-NEC is first diagnosed. It may ease the extremely difficult task parents will have making subsequent decisions regarding the withdrawal of treatment.

It is also critically important to determine the precise length of intestine which remains after bowel resection. This length should not be estimated, but measured using a suture placed along the antimesenteric border of the intestine. The length of remaining bowel is extremely helpful in determining the likelihood of successful intestinal adaptation and the infant's ability to thrive off TPN.

High Jejunostomy in the Setting of Distal Perforation(s)

A high jejunostomy may also be created in pan-NEC when the distal intestine has one or more perforations. There are several treatment options vis-a-vis the remaining distal bowel in this setting. Perforated segments should be resected. If only a single segment is resected, the distal end of the remaining bowel may be brought out as a mucous fistula. In the setting of multiple resections, multiple stomas may be created, keeping in mind that this will reduce bowel length at the time of stoma takedown. If this is of concern, multiple anastomoses stented by a catheter may be created. The infant may then undergo contrast studies at 6-8 weeks as described above. Intestinal continuity may be restored at that time.

The "patch, drain, and wait" technique must be mentioned, since it may be useful in patients with diffuse NEC and multiple areas of perforation. As described, the technique involves simple "patching" of large perforations by simple approximation of open bowel edges, peritoneal drainage with peritoneal drains, and a period of intravenous hyperalimentation. A gastrostomy may also be placed for decompression, and possible feeding.

Specific Considerations: Management of the Very Premature Infant (< 1000 g)

The extremely premature infant who weighs less than 1000 g poses a particular challenge. These infants may not tolerate a laparotomy well. If perforation has occurred, these infants may be treated at the bedside in the neonatal intensive care unit by performing peritoneal drainage using a Penrose drain placed in the right lower quadrant. The technique involves the use of local anesthesia to a small area of the right lower quadrant. When an infant weighing less than 1000 g has been diagnosed as having an intestinal perforation either by the presence of a pneumoperitoneum or by the aspiration of meconium from the peritoneal cavity, this author has used a 1/4 inch Penrose drain tunneled into the right lower quadrant, or brought out through a small counter incision in the left lower quadrant as the initial treatment. Ein reported that no further operative treatment was required in 32% of a group of 41 infants weighing less than 1500 g who were treated with peritoneal drainage; of the remaining infants (68%), these were divided into three groups of roughly equal size: 1) those who died soon after drainage, 2) those who deteriorated and required subsequent laparotomy and 3) those who did well initially, but required delayed operation for bowel obstruction.[7] The authors subsequently published data which showed comparable survival between laparotomy and peritoneal drainage in infants weighing more than 1000 g; in infants less than 1000 g, peritoneal drainage had a clearly higher survival.[8] The general approach of peritoneal drainage in the very premature infant has been advocated by the authors.[9]

Treatment of Post-NEC Strictures

Strictures occur in 10-20% of NEC infants who are managed without surgery, but may also occur in patients who have undergone surgery. A 1995 paper reported a postoperative stricture rate of 9%.[10] Strictures usually occur in bowel segments involved with NEC, and is presumably the result of scarred healing of the involved bowel. The frequency of strictures has increased, presumably as a result of increased overall survival in NEC. The left colon is the most common site of strictures, followed by the right colon and the terminal ileum. All patients who were treated with operation and proximal enterostomy should have a contrast enema performed prior to stoma closure so that strictures can be identified preoperatively. Some clinicians routinely perform contrast enemas in all patients treated medically prior to the resumption of feedings, but the majority do not perform studies unless the patient develops symptoms once feedings have started. Post-NEC strictures in the colon have been treated in some cases with balloon dilatation, but the majority are treated by operative resection and reanastomosis.

References

1. VanderKolk WE, Kurz P, Daniels J et al. Liver hemorrhage during laparotomy in patients with necrotizing enterocolitis. J Pediatr Surg 1996; 31(8):1063-1066.
2. Vaughan WG, Grosfeld JL, West K et al. Avoidance of stomas and delayed anastomosis for bowel necrosis: The "clip and drop-back" technique. J Pediatr Surg 1996; 31(4):542-545.
3. Ade-Ajayi N, Kiely E, Drake D et al. Resection and primary anastomosis in necrotizing enterocolitis. J Royal Soc Med 1996; 89(7):385-388.
4. Musemeche CA, Kosloske AM, Ricketts RR. Enterostomy in necrotizing enterocolitis: An analysis of techniques and timing of closure. J Pediatr Surg 1987; 22:479.
5. Weber TR, Tracy TF Jr, Silen ML et al. Enterostomy and its closure in newborns. J Pediatr Surg 1995; 130(5):534-537.
6. Moore TC. The management of necrotizing enterocolitis by "patch, drain and wait". Pediatr Surg Int 1989; 4:110-113.
7. Ein SH, Shandling B, Wesson D et al. A 13-year experience with peritoneal drainage under local anesthesia for necrotizing enterocolitis perforation. J Pediatr Surg 1990; 25(10):1034-1037.
8. Azarow KS, Ein SH, Shandling B et al. Laparotomy or drain for perforated necrotizing enterocolitis: Who gets what and why? Pediatr Surg Int 1997; 12:137-139.
9. Morgan LJ, Schochat SJ, Hartman GE. Peritoneal drainage as primary management of perforated NEC in the very low birth weight infant. J Pediatr Surg 1994; 29(2):30-34.
10. Horowitz JR, Lally KP, Cheu HW et al. Complications after surgical intervention for necrotizing enterocolitis: A multicenter review. J Pediatr Surg 1995; 30(7):994-998.

CHAPTER 7

The Pathology of Necrotizing Enterocolitis

Katrine Hansen

N ecrotizing enterocolitis is the most common condition for which emergency gastrointestinal surgery is required during the neonatal period. Epidemiologic studies indicate that prematurity is without question the most important risk factor for NEC. There is increased incidence of NEC with decreasing birth weight and gestational age. Like incidence, the age of onset is inversely related to birth weight and gestational age.[1] The severity of the disease and resulting complications are greater in infants of extremely low birthweight with more extensive intestinal involvement and higher mortality.[2] Recent advances in neonatal intensive care have led to increased survival of smaller, more immature infants who are at greatest risk for NEC. As smaller newborns survive early causes of death, NEC mortality may increase.[3]

Pathology

The diagnosis of NEC is a based on clinical and radiologic findings. Pathologic examination, however, of resected bowel segments and autopsy specimens has been the basis for development of theories of pathogenesis.

In general, the most severely affected portions of the gastrointestinal tract are the terminal ileum, cecum, and right colon, although the entire colon and small intestine may be involved by the disease process. The gross appearance of NEC is, frequently, segmental necrosis with intervening "skip" areas, but, as commonly, there is a continuous segment of involvement with circumferential necrosis. The affected areas are distended with thinned walls that may be dark red to black (Fig. 7.1), and if perforation has occurred, there is a gray-green exudate on the peritoneal surface. Pneumatosis intestinalis, if present, is often visible as subserosal gas bubbles (Fig. 7.2). The mucosal surface is hemorrhagic, ulcerated and friable. Perforations are present in approximately 50% of patients.[4,5]

Microscopic features are similar in all cases, but show considerable variation from one field to the next within a single case. Necrotizing enterocolitis in the acute phase is a combination of coagulative-type necrosis and inflammation. The mildest changes consist of necrosis of the mucosa (Fig. 7.3). In more severe involvement, there is transmural necrosis with associated perforation. Hemorrhage and intramural vascular thrombi accompany the necrosis. The thrombi are thought to be secondary to the surrounding injury. Mixed intestinal bacteria are often visible in the lumen or within necrotic superficial mucosa. Submucosal vacuoles, the microscopic counterpart of pneumatosis intestinalis, contain hydrogen, a product of bacterial fermentation (Fig. 7.4). In addition to the evidence of active injury, many cases of NEC show focal reparative epithelial changes and granulation tissue, evidence of healing. These foci of healing within a necrotic bowel suggest that NEC

Necrotizing Enterocolitis, edited by Brian F. Gilchrist. ©2000 Eurekah.com.

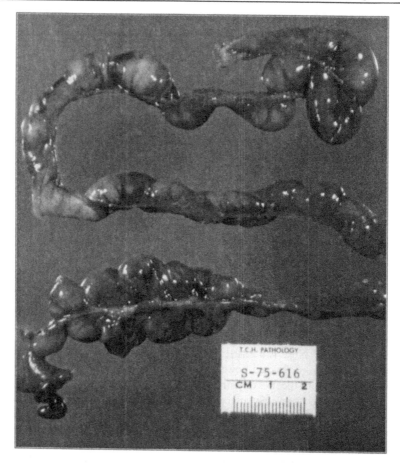

Fig. 7.1. Neonatal necrotizing enterocolitis with multifocal distention and foci of nearly transmural necrosis.

may have a more prolonged course than that which is clinically evident (Fig. 7.5).[4,5] Intestine that has undergone necrosis but is not resected during the acute phase of the disease may develop circumferential submucosal fibrosis during the healing phase. This process manifests clinically as strictures, and these are found in 10-20% of infants 3-10 or more weeks after the diagnosis of NEC.[6]

Pathogenesis

Though the first comprehensive reports of NEC appeared in the literature more than 30 years ago, the pathogenesis of NEC remains incompletely understood. NEC is a disease that predominantly affects preterm infants recovering from various diseases associated with premature birth. There are no consistent identifiable associated disease states that are more commonly noted in patients with NEC compared with gestational age-matched controls. This suggests that NEC is a consequence of, in part, immature intestinal function. It is likely that the primary event in NEC is mucosal injury, but this injury may result from many independent pathways.[7]

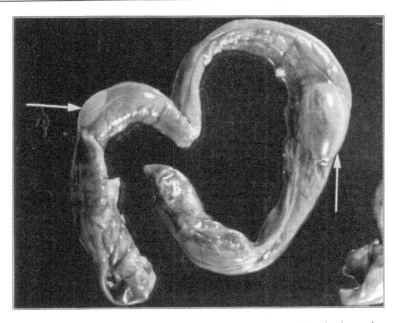

Fig. 7.2. Neonatal necrotizing enterocolitis with advanced pneumatosis intestinalis with subserosal vacuoles seen on the surface of the bowel.

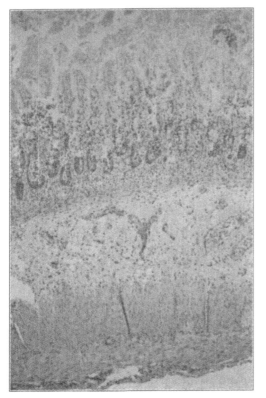

Fig. 7.3. Acute necrotizing enterocolitis with coagulative-type mucosal necrosis (H and E stain, original magnification x 40).

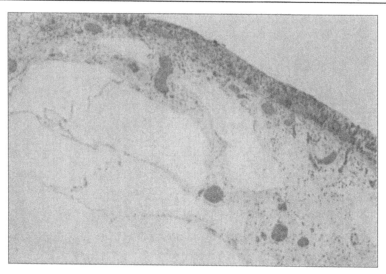

Fig. 7.4. Necrotizing enterocolitis with sloughed superficial mucosa and submucosal pneumatosis intestinalis.

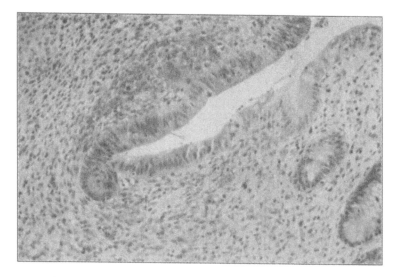

Fig. 7.5. Focus of epithelialization and granulation tissue, evidence of healing, within a bowel with acute necrotizing enterocolitis (H and E stain, original magnification x 200).

Hypoxia/Ischemia

Coagulation necrosis is the dominant histologic finding in NEC. Ischemia is, by far, the most common cause of coagulative necrosis. The ileocolic location of NEC in most cases suggests an ischemic cause because the arteries supplying the ileocolic area are more distal to the origin of the superior mesenteric artery than are arteries supplying more proximal intestine. The site and histology, in addition to the knowledge of a known redistribution of blood from the intestine to more vital organs in diving animals (the "diving reflex") led to an early theory of pathogenesis suggesting circulatory perturbations to be the most important cause of NEC. Clinical and physiologic data linking circulatory disturbances to the etiology of NEC have been disputed, and epidemiologic studies reveal that asphyxia, hypotension and other potentially ischemic events are not risk factors for the development of NEC.[7]

Animal studies, however, have shown a substantial change in the role of intestinal oxygen use during postnatal life. The data show that gut oxidative demand peaks during early newborn life. Associated with this demand is newborn intestinal circulation that appears specifically adapted to meet the increased rate of O_2 use. The higher demand, however, may make it more susceptible to hypoxia. In experimental animals, the first signs of hypoxic intestinal injury are in the intestinal mucosa.

The immature regulation of intestinal blood flow in the neonate is a potential cause of intestinal ischemia.[7] Reflex vasodilation in response to arterial hypotension and hypoxia may be a limiting factor because of the immaturity of the neonatal mammal's intestinal vascular regulation.[8]

Enteral Feedings

More than 90% of infants with NEC have received enteral feedings. Though feeding seems to be related to the onset of NEC, the mechanism connecting the two events is unclear. The volume or the rate of increase in the volume of milk feedings given to premature infants may predispose them to NEC. Overfeeding may create a metabolic demand that is unable to be met by an increase in mesenteric blood flow. It is known that feeding promotes bacterial colonization of the bowel lumen. Ballance et al, in a pathologic study of 84 patients with NEC, found that the degree of bacterial overgrowth in NEC is greater than that observed in cases of ischemic bowel disease alone.[4]

Colonic bacteria ferment lactose to organic acids and hydrogen gas, and it is known that hydrogen is the main component of pneumatosis intestinalis. The production of gas could lead to abdominal distension, which, in association with paralytic ileus, can increase luminal tension. Increased wall tension may lead to compromise of mucosal perfusion with resulting mucosal hypoxic injury.

Additionally, in terms of inflammatory mediators, it has been hypothesized that bacterial proliferation following feeding stimulates endotoxin production and increases the risk of NEC.[9]

It has been suggested that some cases of NEC may be a manifestation of allergic proctocolitis. The features of milk protein allergy include large numbers of eosinophils in the lamina propria and eosinophils as a major inflammatory cell in areas of mucosal damage. Tissue eosinophilia is not a common feature in most cases of surgically treated NEC.

Infection

There is epidemiologic and circumstantial evidence indicating that NEC may be caused by one or more infectious agents. NEC has been reported to occur in clustered epidemics. Epidemics have been reported to be associated with the recovery of single pathogens, such as *E. coli*, *K. pneumoniae*, *Salmonella*, *S. epidermidis*, and viruses such as coronavirus, enterovirus and rotavirus. In addition, personnel who work in NICU often become ill during episodes of NEC. Investigations, however, have not been able to identify any known pathogen during epidemics of NEC. The pathologic changes of NEC are distinctly different from those of infectious enteritis or colitis in older patients. The extensive coagulative necrosis seen in NEC is not present in the enteritis caused by *Shigella, Salmonella* or *Campylobacter jejuni*.

In light of the lack of an identified infectious agent, the possibility has been raised that NEC is caused by a toxin producing organism. It is well known that NEC resembles the disease pigbel caused by the enterotoxin of *C. Perfringens* type C. Pigbel shares a number of characteristics with NEC including the development of necrotic lesions in the gastrointestinal tract and the propensity for intestinal perforation.[10]

Inflammatory Mediators

It is well known that reduced blood flow is not the only means of inducing coagulative necrosis in the intestine. Inflammatory mediators such as platelet-activating factor induce similar lesions in experimental animals. PAF is an endogenous phospholipid mediator produced by inflammatory cells, endothelial cells, platelets and bacteria of the intestinal flora. Systemic administration of PAF induces an immediate and sometimes transient hypotensive response. With larger doses, the shock becomes profound and irreversible intestinal necrosis develops rapidly. The sites within the gastrointestinal tract as well as the microscopic features of the mucosal involvement are very similar to human NEC.[11] In a neonatal rat model of NEC, Caplan showed that PAF was a critical mediator in the development of bowel injury. Local and systemic PAF concentrations are elevated in neonates with NEC, and perhaps as importantly, levels of the PAF degrading enzyme, acetylhydrolase are low in newborns.[12] It is suggested by investigators that PAF causes intestinal injury and deleterious systemic changes via a synergistic action with endogenous bacterial polymer toxins, presumed to be a product of intestinal bacteria. Bacteria produce endotoxins that instigate the inflammatory cascade by activating PAF, TNF, and interleukin-1. A currently prevailing hypothesis suggests that splanchnic hypoperfusion may induce mucosal injury in the gut. Intestinal microorganisms breach the immature neonatal gut barrier made weaker by the ischemic insult. An inflammatory response results in an inflammatory cascade leading to the clinical and histologic entity we know as NEC.[12]

The etiology of NEC is multifactorial. In addition to decreased mesenteric oxygen delivery, bacterial colonization of the gastrointestinal tract is generally held to be an important requisite for the development of NEC. Similar to sepsis and adult respiratory distress syndrome, NEC seems to involve a final common pathway that includes the endogenous production of inflammatory mediators involved in the development of intestinal injury.

References

1. Stoll BJ. Epidemiology of necrotizing enterocolitis. Clin Perinatol 1994; 21:205-218.
2. Rowe MI, Reblock KK, Kurkchubasche AG et al. Necrotizing enterocolitis in the extremely low birth weight infant. J Pediatr Surg 1994; 29:987-991.
3. Hollman RC, Stoll BJ, Clarke MJ et al. The epidemiology of necrotizing enterocolitis infant mortality in the United States. Am J Pub Health 1997; 87:2026-2031.
4. Balance WA, Dahms BB, Shenker N et al. Pathology of neonatal necrotizing enterocolitis: A ten-year experience. J Pediatr 1990; 117:S6-S313.
5. Joshi VV. Common problems in pediatric pathology. New York, Igaku-Shoin, 1994; 114-142.
6. Kosloske AM, Burstein J, Bartow SA. Intestinal obstruction due to colonic stricture following neonatal necrotizing enterocolitis. Ann Surg 1980; 192:202.
7. Kliegman RM. Pathophysiology and epidemiology of necrotizing enterocolitis. In: Polin RA, Fox WW, eds. Fetal and Neonatal Physiology. WB Saunders, 1998: 1425-1432.

8. Nowicki PT, Nankervis CA. The role of the circulation in the pathogenesis of necrotizing enterocolitis. In: Stoll BJ, Kliegman RM, eds. Clinics in Perinatology. WB Saunders 1994; Vol. 21 219-234.

9. Caplan MS, MacKendrick W. Necrotizing enterocolitis: A review of pathogenetic mechanisms and implications for prevention. Pediatric Pathology 1993; 13:357-369.

10. Kliegman, RM, Walker A, Yolken RH. Necrotizing enterocolitis. Research agenda for a disease of unknown etiology and pathogenesis. Clinics in Perinatology 1994; (2)437-455.

11. Gonzalez-Crussi F, Hseuh W. Experimental model of ischemic bowel. Am J Pathol 1983; 112:127-135.

12. Hsueh W, Caplan MS, Tan X et al. Necrotizing enterocolitis of the newborn: pathogenetic concepts in perspective. Pediatric and Developmental Pathology 1998; 117:2-16.

CHAPTER 8

The Nutritional Concerns in Necrotizing Enterocolitis

Carlotta Hample

Short bowel syndrome (SBS) is a chronic state of malabsorption that occurs after substantial resection of the small intestine. Necrotizing enterocolitis is the leading cause of SBS in anatomically normal neonates. Due to the extensive loss of absorptive surface area and subsequent malabsorption, the clinical course of patients is often complicated by diarrhea, metabolic abnormalities and malnutrition. Although it is a functional definition, the degree of symptoms depends on the length, unique function and adaptive capacity of the remaining small intestine.

Physiology of Short Bowel Syndrome

An understanding of the specific regional functions of the small intestine is essential to comprehend the pathophysiology of short bowel syndrome. The jejunum has a large absorptive surface area and a high concentration of digestive enzymes and carrier proteins.[20] For these reasons, most nutrient absorption occurs in the proximal small bowel. The ileum does not have the absorptive capacity of the jejunum; however, the majority of fluid and electrolytes are absorbed here. In addition, the terminal ileum possesses specific receptor sites for the absorption of vitamin B12 and bile salts.

After massive jejunal resection, few alterations in nutrient or fluid absorption occur because the remaining ileum possesses tremendous adaptive capacity. Ileal resection, on the other hand, is often complicated by osmotic diarrhea. Consequently, large amounts of isotonic fluid and salts are lost through the jejunostomy. This results in hypokalemia, hypomagnesemia and dehydration termed "end jejunostomy syndrome". A similar process takes place in patients with continuity of the colon. Substances that cannot be reabsorbed, such as bile salts, unabsorbed carbohydrates and fatty acids interfere with colonic fluid absorption leading to an osmotic diarrhea. Vitamin B12 and bile salt malabsorption is unique to ileal resection. If the bile salt pool becomes depleted, steatorrhea and fat soluble vitamin deficiency develop.

The ileum also plays an important role in regulating intestinal motility. Slowing the gastric emptying and intestinal transit in response to intraluminal fat is referred to as the ileal brake phenomenon. In addition, regulatory hormones that slow transit time, such as peptide YY, are produced in the ileum.

The ileocecal valve serves as a physical barrier that prevents colonic bacteria from entering the small intestine. Once resected, the small bowel is prone to bacterial overgrowth. Bacteria deconjugate bile salts inhibiting formation of micelles resulting in fat malabsorption. The ileocecal valve also influences absorption by regulating the delivery

Necrotizing Enterocolitis, edited by Brian F. Gilchrist. ©2000 Eurekah.com.

rate of luminal contents from the ileum thereby increasing the time nutrients have in contact with the small bowel mucosa.

The colon not only absorbs fluid but also increases nutrient absorption. An intact colon can serve as a source of additional calories by salvaging short chain fatty acids that result from the bacterial breakdown of unabsorbed carbohydrates. The colon also has the ability to regulate transit time by slowing the gastric emptying.

Adaptation

The remaining small intestine adapts to maximize nutrient absorption thereby potentiating tolerance to enteral nutrition. Both functional and structural adaptation of the remnant bowel begins with epithelial hyperplasia. Enteral nutrition is essential for mucosal hyperplasia to occur. In fact, some degree of mucosal atrophy may result if all nutrients are provided parenterally.[20] There are several mechanisms through which this is accomplished, including stimulation by direct contact of the mucosa with intraluminal nutrients and the release of trophic hormones and gastrointestinal secretions.[20]

Various dietary components have demonstrated trophic potential as shown in Table 8.1.

Many hormones have been evaluated as mediators of intestinal adaptation. In a study of eight patients with short bowel syndrome treated by TPN, Byrne demonstrated that a combination of growth hormone, glutamine and a modified diet enhanced nutrient absorption from the remnant bowel after massive resection.[2] Throughout the study, however, the patients were aggressively fed enterally. Therefore, the adaptive changes may have been influenced by nonspecific nutrient stimulation and not solely attributed to growth hormone and glutamine. In a subsequent study of 31 patients with short bowel syndrome treated with TPN, Byrne showed that 40% of patients treated with the same regimen were able to discontinue parenteral nutrition during 1 year of follow up.[3] Scolapio evaluated the use of glutamine, growth hormone and a high carbohydrate, low fat diet in eight patients with short bowel syndrome treated with TPN.[14] Contrary to previous findings, there was no improvement in small bowel morphology, stool losses or macronutrient absorption. The contributions of glutamine, growth hormone and diet to intestinal adaptation in short bowel patients remain unclear (Table 8.2).

The degree of adaptation determines whether a patient will achieve full enteral nutrition. This is strongly influenced by the length of remaining bowel, the presence of an ileocecal valve and the length of residual ileum. In a study of 13 infants with < 38 cm of small bowel, Dorney concluded that as little as 11 cm of jejunoileum with an ileocecal valve or 25 cm without an ileocecal valve was compatible with survival and adaptation.[4] This was supported by other studies in which full enteral nutrition was attained in patients with as little as 10 cm of jejunoileum.[7,8] The presence of an ileocecal valve in these patients with very short residual small bowel appears to be critically important to adaptation. Goulet studied infants with 40 cm small bowel with and without an ileocecal valve. He demonstrated that the time to full intestinal adaptation was strongly influenced by the presence of the ileocecal valve, requiring 18 months with an ileocecal valve versus 45.3 months without an ileocecal valve.[6] Similarly, other studies have shown that patients without an ileocecal valve took three times as long to undergo intestinal adaptation than those with an intact ileocecal valve.[5] The absence of an ileocecal valve prolongs the length of parenteral nutrition therapy required before adaptation occurs but does not determine successful weaning from it.[5]

Table 8.1. The role of nutrients in intestinal adaptation

Nutrient	Action	Significance
Pectin/fiber polysaccharide	Prevents mucosal atrophy	Metabolized to short chain fatty acids an enterocyte nutrient
Long chain triglycerides	Trophic to mucosa, increase mucosal mass	Maximal effect in mid-small bowel
Medium chain triglycerides	Trophic to mucosa, increase mucosal mass	More effective absorption, better calorie source
Glucose/fructose/mannose	Stimulate mucosal growth	Direct effect
Casein	Stimulates mucosal growth	Greater effect than hydrolase

Table 8.2. The roles of other hormones on intestinal adaptation

Hormone	Action	Significance
Glutamine	Essential for nucleic acid synthesis and cell proliferation, serves as a fuel for enterocytes	Maintains mucosal integrity and prevents bacterial translocation
Enteroglucagon	Stimulates production of polyamines which increases RNA and DNA production	Cellular/mucosal hyperplasia Stimulates adaptation
Growth hormone	Mediated through insulin-like growth factor	Enhances mucosal hypertrophy/hyperplasia
Epidermal growth factor	Increases RNA and DNA synthesis leading to increased cell replication	Unclear
Gastrin	Trophic only in stomach and duodenum	Plays no role in adaptation
Pancreaticobiliary	Increases DNA synthesis and protein concentration	Villus hyperplasia, increased mucosal mass

Nutrition

Controversy remains over the ideal diet to use in patients with short bowel syndrome. A variety of diets have been studied to determine which promotes optimal intestinal absorption. It is generally believed that a diet low in fat but high in carbohydrates and specific fibers will result in less diarrhea and better absorption of nutrients.[22] The premise behind these dietary modifications is that undigested fiber and carbohydrates are metabolized to short chain fatty acids in the colon providing an additional source of calories. Furthermore, fiber slows the intestinal transit time. Nordgaard has recently demonstrated that a high carbohydrate, low fat diet provided for four days can decrease the fecal loss of calories in short bowel patients with normal colon.[13]

Contrary to these findings, several authors have shown that restricting dietary fat does not significantly influence nutrient absorption. Woolf evaluated patients with stable disease and showed that despite fat malabsorption, the total calories absorbed and excreted were the same on a high fat diet as compared to a low fat diet. It was concluded that a low fat diet had no special benefit in the overall nutrition of SBS patients.[22] Other authors have shown that both the total calories absorbed and the volume of output is comparable for both a high carbohydrate and high fat diet.[1]

An elemental formula, which is predigested, would seem to be the ideal diet in the early adaptive changes following small intestinal resection. However, these formulas are hyperosmolar and may induce an osmotic diarrhea. McIntyre, in a comparative study, showed that an elemental diet had no benefit over a polymeric preparation in the short gut syndrome.[12] Polymeric formulas are well tolerated even in the early stages after resection and may contribute to intestinal adaptation.

Clinical Management

Once the immediate postoperative ileus has resolved, large fluid losses from the nasogastric tube and ostomies complicate fluid and electrolyte management. High sodium losses are common. Solutions containing at least 80-100 meq/l of sodium are often required to maintain fluid and electrolyte homeostasis.[20] Adjustments in fluids need to be made frequently. Therefore, a separate replacement solution is beneficial when TPN is in use. TPN used during this time period bridges the gap until adaptation is maximized and enteral feeds are tolerated. TPN is initially given over 24 hours until blood sugars and electrolytes are stable. It can then be cycled overnight with enteral feedings given continuously or as boluses during the day. To prevent fluctuations of blood sugars in small infants, the TPN rate should be tapered up and down when beginning and ending the cycle. As the volume of enteral feeds is increased, the amount of TPN should be decreased to maintain the same amount of caloric delivery. Some patients may not achieve full enteral feeds and remain dependent of TPN for prolonged periods of time.

Postoperatively when bowel sounds are present and the ileus resolves, enteral feedings should be initiated with a continuous infusion. Feeds should be started at a slow rate of infusion and reduced concentration. The rate should gradually be advanced to prevent intolerance. The concentration can be more rapidly advanced to 20 cals/oz as tolerated. To provide the maximal number of calories without excess fluid, the concentration is advanced first. Although concentrating the formula past 20 cals/oz will provide extra calories, it may lead to carbohydrate malabsorption. Evidence of carbohydrate malabsorption includes stool pH < 5.5, the presence of reducing substances or large volume of stool output. Advancing feeds in the face of carbohydrate malabsorption will result in an osmotic diarrhea with excessive fluid losses. An increase in stool losses by more than 50% in a 24 hour period or stool losses greater than 40-50 ml/kg/day suggests osmotic diarrhea, intercurrent illness or that the patient's absorptive capacity is already maximized.[20] Small oral feeds should be instituted early. This will stimulate a suck and swallow and prevent feeding refusal behavior.

Complications

Nutritional: Numerous chronic problems arise during the long term management of a patient with short bowel syndrome. Nutritional deficiencies are common following small bowel resection. As a consequence of steatorrhea, deficiencies of the fat soluble vitamins A, D, E and K are frequent. Since vitamins D and K can be synthesized by the skin and colonic bacterial flora respectively, symptoms due to deficiency are unusual. Neurological sequelae resulting from vitamin E deficiency are largely irreversible but can be prevented with adequate supplementation.[16] However, this may prove difficult since most vitamin E supplements are also fat soluble and therefore poorly absorbed. The use of newer water soluble vitamin E preparation, such as Liqui E, aids in the intestinal absorption of vitamin E in short bowel syndrome.

Divalent cations bind to fatty acids lost in the stool and deficiencies of zinc, calcium and magnesium may develop. Zinc deficiency, referred to as acrodermatitis enteropathica, results in hair loss and rash in the oral, genital and anal areas. This can be prevented or reversed with oral supplementation of zinc. Similarly, calcium deficiency readily responds to oral supplementation of vitamin D and calcium. Magnesium deficiency is more problematic to manage since oral supplementation often leads to an osmotic diarrhea. After ileal resection, patients are prone to vitamin B12 deficiency. Levels should be monitored periodically since a deficiency may take years to develop.

Diarrhea

The majority of patients with short bowel syndrome suffer from watery diarrhea. There are various factors which contribute to excessive intestinal fluid secretions. Unabsorbed carbohydrates generate short chain fatty acids, which can lead to an osmotic diarrhea. Opiates, such as loperamide, improve the diarrhea by slowing intestinal transit time. Codeine is also quite effective, however, it has significant CNS side effects. After ileal resection, unabsorbed bile salts enter the colon causing mucosal irritation and a secretory diarrhea. This is referred to as choleretic diarrhea and is best managed with cholestyramine or other exchange resins that bind unabsorbed bile salts.

Hypergastrinemia frequently follows resection and is known to cause a secretory diarrhea. In cases where cimetidine is ineffective, omeprazole, a proton pump inhibitor, may be useful. Somatostatin, which decreases the production of gastrointestinal secretions and slows transit time, has been used in a limited number of patients. The limited experience using this drug has provided varying results and its use is not clearly defined.

Bacterial Overgrowth

Colonization of the small intestine is a frequent occurrence in the presence of altered peristalsis and resection of the ileocecal valve. Bacteria deconjugate bile salts leading to bile salt deficiency which promotes steatorrhea and malabsorption of fat soluble vitamins. In addition, bacteria alter mucosal absorption by causing inflammatory damage and directly compete for nutrients such vitamin B12. Clinically, patients develop abdominal cramping, bloating, diarrhea and gastrointestinal blood loss. A diagnosis is made by aspiration and culture of the small bowel secretions. Alternatively, breath hydrogen testing can be performed. A rapid rise in the fasting breath hydrogen following ingestion of glucose (2g/kg max of 50 g) is suggestive of bacterial overgrowth. False positives are likely to occur if the transit time of the small intestine is so rapid that the unabsorbed glucose is transported immediately into the colon. Antibiotics given for the first five days of a month should provide adequate treatment. Commonly used antibiotics include Flagyl, Bactrim and oral gentamycin.

Colitis

Short bowel colitis is a noninfectious colitis characterized by bloody, watery stools that develop when enteral feeds are advanced. The symptoms cease when feedings are withheld suggesting that malabsorbed nutrients may be responsible. Colonic biopsy specimens reveal edema, patchy erythema, friability, lymphoid hypertrophy and eosinophilic infiltrate.[17] Treatment with sulfasalazine usually allows feeds to be advanced without further symptoms. Occasionally, antibiotics and steroids and required.

Nephrolithiasis

Patients with ileal resection and an intact colon are predisposed to develop kidney stones. In patients with steatorrhea, calcium preferentially binds to intraluminal free fatty acids and is therefore not available to form nonabsorbable calcium oxalate. This leads to increased colonic absorption of oxalate, which subsequently precipitates with calcium in the kidney to form stones. Urinary oxalate should be monitored routinely. Treatment begins with calcium supplementation and a low oxalate diet. This will promote calcium oxalate binding in the colon. In addition, cholestyramine will bind oxalate in the colonic lumen and decrease its absorption.

TPN Related Complications

Although TPN allows good development and quality of life, it is associated with considerable morbidity. Hepatobiliary complications, including cholestatic jaundice and cholelithiasis, occur frequently. In fact, liver disease due to TPN accounts for one third of deaths in patients receiving long term TPN.[18] The use of amino acid solutions specifically formulated for infants and children, aggressive treatment of catheter related sepsis and prevention of small bowel bacterial overgrowth significantly decreases the incidence of cholestatic jaundice. The early institution of enteral feeds is important in protecting patients from cholestasis. Providing at least 23-30% of total daily caloric intake as enteral feeds plays a significant role in preventing TPN related liver disease.[20] TPN related cholelithiasis is associated with gallbladder stasis, malabsorption of bile salts and altered bilirubin metabolism. Other factors contributing to stasis and sludge formation include decreased oral intake and medications such as an anticholinergics and narcotic analgesics. Patients maintained on long term TPN should have periodic ultrasounds to evaluate for biliary disease. It has been recommended that patients receiving long term TPN undergo prophylactic cholecystectomy.[11] (The Editor strongly recommends a non-operative approach in this setting.) Finally, complications of central venous catheter necessary for long term and home TPN have been well documented and include sepsis, thrombosis or central veins and extravascular infiltration.[21]

Intestinal Transplant

Currently transplant is not recommended for infants and children without liver disease who are easily managed on parenteral nutrition but is a potential lifesaving option for those patients whose total small intestinal mass, even after adaptation, will not support life. At present the best candidates for small bowel or liver-small bowel transplant are patients who are permanently dependent on TPN and who have significant TPN-related complications such as loss of vascular access, recurrent episodes of sepsis or hepatic failure.[15] Most transplants performed in children have been combined liver-small bowel due to severe TPN induced liver disease. When the liver is transplanted simultaneously with other organs, it has a protective effect by inducing donor-specific tolerance. This reduces the risk of rejection and prolongs the survival and function of other transplanted organs. The survival rate for combined liver-small bowel transplant is 65%.[10] Isolated small intestinal transplant prior to the development of liver disease is more controversial. Major obstacles to successful small bowel transplantation include rejection, infection and posttransplant lymphoproliferative disease. The large quantity of lymphoid tissue present in the intestinal graft increases the risk of graft versus host disease. However, the use of more potent immunosuppressant agent, such as FK 506, significantly increased the survival rate and decreased the incidence of rejection. The transplant groups at the University of Pittsburgh and the University of Nebraska Medical Center have successfully undertaken isolated small intestinal transplantation with the use of FK 506 for immunosuppression.[9, 19] Many of the complications following transplantation are related to the intense immunosuppression used to prevent rejection. Associated with the use of these immunosuppressive agents is a surge of viral infections, such as CMV and EBV, and posttransplant lymphoproliferative disease (an EBV-driven malignancy). These complications may respond to a decrease or cessation of the immunosuppression but this increases the chance of rejection. Intestinal transplant is evolving as a promising alternative for patients with chronic intestinal failure and complications of long term TPN support.[18] With diligence, effective techniques probably can be established to reduce the risk of rejection, lymphoproliferative disease and infection to a level at which elective transplantation of the small bowel can become a reality.[20]

References

1. Vanderhoff J. Short bowel syndrome in children and small intestinal transplant. Ped Clin N Amer 1996; 43:533-549.
2. Byrne TA, Morrissey TB, Nattakom TV et al. Growth hormone, glutamine and a modified diet enhance nutrient absorption in patients with severe bowel syndrome. JPEN 1995; 19:296-302.
3. Byrne TA, Persinger RL, Young LS et al. A new treatment for patients with short bowel syndrome. Ann Surg 1995; 222:243-255.
4. Scolapio JS, Camilleri M, Fleming CR et al. Effect of growth hormone, glutamine and diet on adaptation in short bowel syndrome: A randomized, controlled study. Gastroenterology 1997; 113:1074-1081.
5. Dorney SFA, Amet ME, Berquist WE et al. Improved survival in very short bowel with use of long-term parenteral nutrition. J Ped 1985; 107:521-524.
6. Iacono G, Carroccio A, Cavataio F et al. Extreme short bowel syndrome: a case for reviewing the guidelines for predicting survival. J Ped Gastroenterology Nutr 1993; 16:216-219.
7. Kurkchubasche AG, Rowe MI, Smith SD. Adaptation in short bowel syndrome: reassessing old limits. J Ped Surg 1993; 28:1069-1071.
8. Goulet OJ, Revillon Y, Jan D et al. Neonatal short bowel syndrome. J Ped 1985; 107:521-524.
9. Georgeson KE, Breaux CW. Outcome and intestinal adaptation in neonatal short bowel syndrome. J Ped Surg 1992; 27:344-350.
10. Woolf GM, Miller C, Kurian R. Diet for patients with a short bowel: High fat or high carbohydrate? Gastroenterology 1983; 84:823-828.
11. Nordgaard I, Hansen BS, Mortensen PB. Colon as a digestive organ in patients with short bowel. Lancet 1994; 343:373-376.
12. Allard JP, Jeejeebhoy KN. Nutritional support and therapy in the short bowel syndrome. Gastro Clin N Amer 1989; 18:589-591.
13. McIntyre PB. The short bowel. Br J Surg 1985; 72:592-593.
14. Sokol RJ. Vitamin E deficiency and neurological disorders. In: Packer L, Fucks J, eds. Vitamin E

CHAPTER 9

The Use of Small Bowel Transplantation in Necrotizing Enterocolitis

George Tsoulfas and Christopher Breuer

S hort bowel syndrome (SBS) is the most serious long-term gastrointestinal com plication associated with surgically treated NEC. SBS occurs in up to 23% of survivors who undergo surgical management for NEC.[1] Patients who develop SBS require parenteral nutrition for survival. The long-term use of parenteral nutrition in this patient population is associated with the development of cholestasis which can lead to cirrhosis, sepsis, and increased mortality. In patients with sustained direct bilirubin levels of more than 4 mg/dl for more than 6 months, an 80% mortality rate is expected.[2] Additionally the financial burden of prolonged parenteral nutrition use is heavy. The average cost of maintaining a pediatric patient on home parenteral nutrition ranges from $100,000 to $150,000 per year.[2] Fortunately most patients with SBS eventually undergo sufficient intestinal adaptation to be weaned off parenteral nutrition. Long-term survival in children with SBS is expected to be at least 85%.[3] In those patients with SBS secondary to NEC that are refractory to medical or surgical management, bowel transplantation remains an option.

Necrotizing Enterocolitis

When the diagnosis of NEC is suspected, supportive treatment should be immediately initiated. The principles of medical management include: cardiovascular resuscitation and support, work-up and control of sepsis, along with close observation for signs of intestinal gangrene or perforation. Using medical management alone, one half to two thirds of infants with NEC will recover fully. The remaining patients will require surgical intervention. Those who require operation are the sickest infants with advanced NEC[4] with reported case fatality rates between 20-40%.[5-6] The principles guiding surgical interventions for NEC are designed to prevent SBS. These include excision of gangrenous bowel, exteriorization of marginally viable bowel, and preservation of as much intestinal length as possible.[4]

Most pediatric surgeons are reluctant to operate without clear indications of intestinal gangrene because a negative exploratory laparotomy is not an acceptable diagnostic procedure for a critically ill premature infant.[4] Ideally surgery should be performed after the onset of intestinal gangrene, but before perforation, in order to reduce the operative mortality.[7] Currently massive intestinal resection leaving insufficient bowel length to provide adequate bowel function is not recommended.

Massive ischemic injury involving the entire small and large intestine is referred to as NEC Totalis and is found in 18-38% of operations for NEC.[8] Surgical options in this situation include: open and close laparotomy, proximal diversion with second look surgical exploration, or massive resection . Long-term survival after NEC Totalis is limited, resulting in short bowel syndrome in the majority of survivors regardless of the surgical option selected.

Short Bowel Syndrome

Short bowel syndrome (SBS) is defined as malabsorption, fluid and electrolyte loss and malnutrition after massive resection of the small intestine.[9] Wilmore retrospectively reviewed the case histories of 50 infants younger than 2 months of age and defined SBS as less than 75 cm of residual small intestine or less than 40% of the normal length of the small intestine in a full-term neonate.[10] More recent studies have documented return to full enteral feeding in patients with as little as 10 cm of remaining small bowel.[11] As much as half of the small bowel can be resected without developing SBS provided the duodenum, distal ileum and ileocecal valve are spared. In contrast, distal ileal resections that include the ileocecal valve can cause severe diarrhea even though only 25% of the small intestine has been resected.[9] The increased susceptibility to SBS associated with ileal resection is due to several factors including the ileum's capacity for compensatory hyperplasia, the ileum's unique concentrating abilities, and the importance of the ileocecal valve in preventing bacterial overgrowth. Unfortunately, NEC most commonly involves the ileocecal region accounting for 44% of cases of NEC requiring surgical resection.[12] In fact NEC is one of the most common causes of SBS accounting for up to 46% of the cases of SBS in infants.[13]

In addition to the amount of intestine resected, the condition of the remaining bowel plays an important role in the development of SBS. Since a major guiding principle in NEC surgery pertains to bowel length preservation, sections of viable but damaged bowel can be preserved. The long-term effects of ischemic damage from NEC on these bowel segments include malabsorption, dysmotility and the formation of strictures, which can cause partial obstruction leading to bacterial overgrowth, which further compounds the problems associated with SBS. Specific clinical consequences of SBS include: nutrient malabsorption, vitamin deficiency, electrolyte losses and trace elements deficiency. In addition SBS causes gastric acid hypersecretion and lactic acidosis.

The initial treatment of SBS is supportive. The use of total parenteral nutrition (TPN) has led to the prolonged survival and eventual recovery in most patients with SBS. Even in patients who ultimately go on to full recovery with resumption of complete enteral feeding, TPN provides a bridge to recovery by allowing full nutritional support during the months to years required for intestinal adaptation. The most important component in the treatment of SBS is the early resumption of trophic feeding which is essential for intestinal adaptation. Other important adjuncts in the treatment of SBS include: the use of H2 blockers to control the gastric hypersecretion, maintenance of adequate fluid balance, supplementation of deficient electrolytes and vitamins, and the use of bile acid replacement. The periodic use of oral antibiotics such as: Trimethoprim sulfate, Metronidazole and Neomycin in an effort to prevent bacterial overgrowth can be important adjuncts to SBS care.[14] Studies are currently underway evaluating the use of growth factors including the use of glutamine and growth hormone supplementation as means to promote intestinal adaptation.[15]

A major limiting factor regarding the medical management of SBS relates to the complications of long-term use of parenteral nutrition. TPN use is associated with liver injury in neonates, especially low birth weight premature infants who are most susceptible to NEC. Mild elevation of transaminases and conjugated bilirubin levels may be observed within 2 weeks of initiating TPN.[16] Long-term use of TPN causes cholestasis which can lead to the development of cirrhosis, portal hypertension and ultimately liver failure. Additionally TPN requires maintenance of long-term central venous access which can lead to multiple episodes of line sepsis or other line-related complications such as central vein thrombosis or loss of central venous access.

Multiple surgical procedures have been proposed in order to alleviate SBS. Most procedures involve either performing a bowel lengthening procedure such as the Bianchi procedure[17] or slowing down intestinal transit time by creating a valve or inserting an antiperistaltic segment of bowel or colon.[15] None have met with uniform success, although some patients have been able to be weaned off of parenteral nutrition using these surgical techniques. More importantly surgery plays an major role in relieving partial bowel obstructions caused by adhesions or strictures secondary to NEC. Partial obstruction leads to bacterial overgrowth which can worsen the effects of SBS by causing dumping. Prompt surgical correction of partial bowel obstruction is a vital component in the proper management of SBS.

Intestinal Transplantation

Patients with SBS who lose central venous access or who develop hyperalimentation-induced liver disease need to undergo intestinal transplantation. Despite more than 30 years of clinical attempts at intestinal transplantation, it still remains an experimental therapy.[18] The evolution of intestinal transplantation parallels the development of immunosuppressive medications. Before the advent of cyclosporine, multiple attempts at intestinal transplantation using immunosuppressive regimens such as azathioprine, corticosteroids and antilymphocytic antibodies were performed without long-term survivors. The Ontario group performed the first successful liver and bowel transplantation using cyclosporine immunotherapy in 1988.[19] The advent of tacrolimus (FK 506) ushered in the modern era of bowel transplantation.

According to the International Intestinal Transplant Registry,[20] between 1990 and 1995 there were 180 small bowel transplantations from 24 different programs. Two thirds of the recipients were children, with the main indication (64%) being short bowel syndrome. The Pittsburgh small bowel transplantation experience[21] is the most extensive series to date. Between 1990 and 1995, 41 children received 44 transplants under a regimen of tacrolimus and steroids. These included isolated small bowel (SB) (n=10), liver and small bowel (LSB) (n=27) and mutlivisceral (MV) (n=7). The major etiology was short bowel syndrome secondary to extensive resection, necrotizing enterocolitis being the primary etiology in 6 out of the 41 patients. Of the nine SB graft recipients, eight (88.9%) are alive 0.5 to 48 months posttransplantation. The LSB and MV composite grafts shared similar patient (50%) and graft survivals (48.2% and 42.9% respectively). TPN was discontinued an average of 56 days posttransplantation. Twenty children had functioning grafts and were independent of TPN. Some interesting observations include the fact that SB alone appeared at least in the short term to be more successful than LSB or MV, something which is contrary to the earlier transplantation doctrine of the liver graft increasing tolerogenicity. This could be because these patients are healthier than the ones who were also in need of a hepatic graft, or because of lesser immunosuppressive requirements, or simply because it may be technically easier. However, it should be stressed that no long term follow-up is yet available.

The University of Nebraska Medical Center[22] also has made a significant contribution to the field of intestinal transplantation with 27 transplants in 26 infants and children. Ten of these were isolated SB and the rest were LSB. Approximately 50% of the transplants had short bowel syndrome as the etiology. The first of these procedures were done with cyclosporine A, the remainder were with tacrolimus. One year patient survival was 100% for isolated SB and 65% for LSB. This series adds more support to the idea that patients could possibly benefit from isolated small bowel transplantation prior to the onset of endstage liver disease.

The major problems surrounding transplantation include donor organ scarcity, rejection, infection and malignancy. Donor organ scarcity refers to the limited number of cadaveric organs available for transplantation. Despite the ever increasing indications for transplantation (i.e., SBS as an indication for intestinal transplantation), the number of donor organs has remained relatively fixed. Current waiting times for patients listed for intestinal transplantation range from 1-2 years. Many patients die prior to transplantation due to the prolonged waiting periods. The use of living related donors and reduced size grafting has improved the problem; however major social changes designed to increase organ donation will be required to ultimately overcome this problem.

Problems with rejection, infection and malignancy relate to the currently used immunosuppressive regimens, according to the Pittsburgh intestinal transplant experience, between 1990-1996. Histologically proven rejection occurs in 85% of the recipients with an incidence of 2.6 episodes per graft[23] while using tacrolimus immunosuppression. Infections occurred in all transplant patients despite prophylactic use of antibiotics and accounted for nearly 15% of all deaths.

The development of lymphoproliferative disorders is the major long-term problem associated with intestinal transplantation occurring in nearly 30% of patients within 8 months of transplantation.[24] The administration of OKT3 is a major risk factor for the development of postoperative lymphoproliferative disorders. The enormous financial cost and use of limited resources must be factored into the decision making process. The average cost for an intestinal transplantation ranges between $500,000-$600,00 for the transplantation alone. The cost associated with long-term management of the immuno-suppressed individual is astronomical. In this era of managed care these expenses will be closely scrutinized.

Conclusion

Despite major advances in the field of transplantation including the encouraging early results for the bowel transplantation in infants and children suffering from SBS secondary to NEC, this modality is still experimental. Long-term follow-up is needed to better define the problems associated with this therapy including issues regarding rejection, infection and malignancy. Major advances must still be made before this modality can be considered the standard of care. Until these advances have been made, bowel transplantation should only be used on a compassionate needs basis for those patients with no other hope for long-term survival.

References

1. Ricketts RR. Surgical treatment of necrotizing enterocolitis and the short bowel syndrome. Clinics Perinatol 1994; 21:365-370.
2. Teitelbaum DH, Coran AG. Nutrition in pediatric surgery. In: O'Neill JA, Rowe MI, Grosfeld JL, Funkalsrud EW, Coran AG, ed. Pediatric Surgery. 5th ed, Boston: Mosby Year Book, 1998; 171-196.

3. Georgeson KE. Short bowel syndrome in pediatric surgery. In: O'Neill JA, Rowe MI, Grosfeld JL, Funkalsrud EW, Coran AG, ed. Pediatric Surgery. 5th ed, Boston: Mosby Year Book, 1998; 1223-1232.

4. Colossi AM. Necrotizing enterocolitis. In: Oldham KT, Colombani PM, Foglia RP, ed. Surgery of Infants and Children: Scientific Principles and Practice. Philadelphia: Lippincott-Raven Publishers. 1997; 1201-121.

5. Holman RC, Stehr-Green JK, Zelasky MT. Necrotizing enterocolitis mortality in the United States. AM Journal Public Health 1989; 79-987.

6. Kliegman RM, Fanaroff AA. Necrotizing enterocolitis. NEJM 1984; 310:1093-1098.

7. Koloske AM, Papile CA, Burnstein J. Indications for operation in acute necrotizing enterocolitis of the neonate. Surgery 1980; 87:502-507.

8. Koloske AM. A unifying hypothesis for pathogenesis and prevention of necrotizing enterocolitis. Pediatric 1990; 117:568-573.

9. Treem WR. Small intestine. In: Oldham KT, Colombani PM, Foglia RP, ed. Surgery of Infants and Children. Scientific Principles and Practice. Lippincott-Raven Publishers. 1997; 1163-1180.

10. Wilmore DW. Factors correlating with a successful outcome following extensive intestinal resection in newborn infants. J Pediatr 1972; 80:88-93.

11. Kurkchubasche AG, Rowe MI, Smith SD. Adaptation in short-bowel syndrome: Reassessing old limits. J Pediatr Surg 1993; 28(8):1069-1071.

12. Ballance WA, Dakins BB, Sheuher N et al. Pathology of neonatal necrotizing enterocolitis. J Pediatr 1990; 117:56-61.

13. Grosfeld JL, Rescola FJ, West JW. Short bowel syndrome in infants and children. AM J Surg 1986; 151:41-46.

14. Allard JP, Jeejeebhoy KN. Nutritional support and therapy in the short bowel syndrome. Gastroenterol 1989; 18:589-594.

15. Burne TA, Morrissey TB, Ziegler TR et al. Growth hormone, glutamine and fiber-enhanced adaptation of remnant bowel following massive intestinal resection. Surgical Forum 1992; 43:151-156.

16. Quigley EMM, Marsh MN, Shaffer JL et al. Hepatobiliary complication of total parental nutrition. Gastroenterol 1993; 104:286-291.

17. Thompson JS. Surgical aspects of short-bowel syndrome. AM J Surg 1995; 170:532-536.

18. Colombani PM, Lund D. Uncommon and experimental pediatric transplant procedure. In: Oldham KT, Colombani PM, Foglia RP, ed. Surgery of Infants and Children. Scientific Principles and Practice. Philadelphia: Lippincott-Raven Publishers. 1997; 769-782.

19. Grant D, Wall W, Mimeault R et al. Successful small bowel liver transplantation. Lancet 1990; 335:18-23.

20. Grant DR. Current results of intestinal transplantation. Lancet 1996; 347:1801-1803.

21. Todo S, Reyes J, Furukawa H et al. Outcome analysis of 71 clinical intestinal transplantations. Ann Surg 1995; 222(3):270-282.

22. Tzakis AG, Nery JR, Thompson J et al. New immunosuppressive regimens in clinical intestinal transplantation. Transplant Proc 1997; 29:683-685.

23. Kocoshis SA. Small bowel transplantation in infants and children. Gastroenterol Clin North AM 1994; 23(4):727-742.

24. Todo. Intestinal transplantation. In: O'Neill JA, Rowe MI, Grosfeld JL, Funkalsrud EW, Coran AG, ed. Pediatric Surgery. 5th ed. Boston: Mosby Year Book, Inc. 1998; 605-612.

AFTERWORD

The answer to necrotizing enterocolitis lies in its study, not in post facto observation. It is apparent from these pages that early warning is needed; a marker of prognostic and potential significance needs to be discerned. NEC, however, insidiously assaults the infant. If a marker can be found before the cascades are unleashed, then we can meet NEC on a level battlefield. I am confident that just as Survanta conquered hyaline membrane disease, there will be an intervention utilized to protect the neonatal gut from the incitements of NEC. Survanta took premature lung parenchyma and transformed it into functioning, respiring tissue; so too will the infant GI tract be addressed in the future. Whether an interferon or an inducer of specific T-cells or a nutrient will be the therapy remains to be seen. However, we will arrive at the answer and so diminish the slaughter that NEC renders.

We at Downstate, and many others around the world, are studying immunologic parameters in newborns. Our hope is to identify cell subsets that can confer immunologic protection to the immature GI endothelium. It is my belief that in the future we will be able to deliver to the GI tract either enterally or parenterally, or in combination, the protection necessary to maintain the integrity of the GI system. Should we be able to manipulate this interior milieu, we will not have to answer the retorts of the next generation that might have asked why we did not block and treat NEC. We will succeed if our focus is on looking at the disease before it delivers its ravages. We have argued here for a paradigm shift in thinking, as we intend to fight the disease before it gains the high ground. Hold the high ground!

Brian F. Gilchrist, M.D., F.A.C.S.

Index